SpringerBriefs in Statistics

For further volumes:
http://www.springer.com/series/8921

Bernt J. Leira

Optimal Stochastic Control Schemes Within a Structural Reliability Framework

 Springer

Bernt J. Leira
Department of Marine Technology
Norwegian University of Science and Technology
Trondheim
Norway

ISSN 2191-544X   ISSN 2191-5458   (electronic)
ISBN 978-3-319-01404-3   ISBN 978-3-319-01405-0   (eBook)
DOI 10.1007/978-3-319-01405-0
Springer Cham Heidelberg New York Dordrecht London

Library of Congress Control Number: 2013944535

© The Author(s) 2013
This work is subject to copyright. All rights are reserved by the Publisher, whether the whole or part of
the material is concerned, specifically the rights of translation, reprinting, reuse of illustrations,
recitation, broadcasting, reproduction on microfilms or in any other physical way, and transmission or
information storage and retrieval, electronic adaptation, computer software, or by similar or dissimilar
methodology now known or hereafter developed. Exempted from this legal reservation are brief
excerpts in connection with reviews or scholarly analysis or material supplied specifically for the
purpose of being entered and executed on a computer system, for exclusive use by the purchaser of the
work. Duplication of this publication or parts thereof is permitted only under the provisions of
the Copyright Law of the Publisher's location, in its current version, and permission for use must
always be obtained from Springer. Permissions for use may be obtained through RightsLink at the
Copyright Clearance Center. Violations are liable to prosecution under the respective Copyright Law.
The use of general descriptive names, registered names, trademarks, service marks, etc. in this
publication does not imply, even in the absence of a specific statement, that such names are exempt
from the relevant protective laws and regulations and therefore free for general use.
While the advice and information in this book are believed to be true and accurate at the date of
publication, neither the authors nor the editors nor the publisher can accept any legal responsibility for
any errors or omissions that may be made. The publisher makes no warranty, express or implied, with
respect to the material contained herein.

Printed on acid-free paper

Springer is part of Springer Science+Business Media (www.springer.com)

Contents

1   **Introduction**   1
    References   2

2   **Structural Limit States and Reliability Measures**   3
    2.1   Introduction   3
    2.2   Failure Function and Probability of Failure.   4
    2.3   Time-Varying Load and Capacity   7
    2.4   Simplified Calculation of Failure Probability
         for Gaussian Random Variables   9
    2.5   Calculation of Failure Probability for Non-Gaussian
         Random Variables   10
    References   12

3   **Dynamic Structural Response Analysis and Probabilistic**
    Representation   13
    3.1   General   13
    3.2   Environmental Modeling   14
    3.3   Dynamic Response Analysis   16
    3.4   Response Statistics for Scalar Gaussian Processes   17
    3.5   Reliability Formulations for Scalar Gaussian Processes   22
    3.6   Multi-Dimensional Response Statistics   24
         3.6.1   Introduction   24
         3.6.2   Outcrossing-Rate Analysis   25
         3.6.3   Local Maxima and Extremes   26
         3.6.4   Fatigue Analysis for Multi-Dimensional Processes   30
    References   32

4   **Categories of On-Line Control Schemes Based**
    on Structural Reliability Criteria   35
    4.1   General   35
    4.2   Control Schemes Involving Structural Reliability Criteria   36
         4.2.1   Introduction   36
         4.2.2   Various Types of Simplified Structural
              Reliability Indices   36

| | | 4.2.3 | Simplistic Illustrative Example and Comparison of Reliability Indices | 40 |

4.2.3   Simplistic Illustrative Example and Comparison
        of Reliability Indices 40
4.2.4   Off-Line Control Schemes 42
4.2.5   Control Scheme Based on Reliability Monitoring 43
4.2.6   On-Line Control Schemes 44
4.3   Off-Line "Calibration" of LQG Schemes 46
4.3.1   General 46
4.3.2   LQG Control of Stationary Quasi-Static Response 46
4.3.3   Alternative Loss Function Based on Structural
        Failure Cost 48
4.3.4   The Case with Two-Scale Response Characteristics 52
4.3.5   Application to Simplified Example 54
4.4   Concluding Remarks 55
References 56

5   Example Applications Related to On-Line Control Schemes 57
5.1   General 57
5.2   Vessel Dynamics 58
5.2.1   Vessel Motion 58
5.2.2   PID Control Algorithm for Dynamic Positioning 60
5.3   Example of Control Scheme Based on Structural
      Reliability Monitoring 61
5.3.1   General 61
5.3.2   System Model and Behavior Without Position Control .. 61
5.3.3   System Model and Behavior with Position
        Control Implemented 64
5.4   Control Schemes Based on On-Line Computation
      of Reliability Measures 70
5.4.1   Example 1 70
5.4.2   Example 2: Position Mooring of Floating
        Vessel Based on Reliability Index Criteria 87
5.4.3   Further Examples of Application 91
5.5   Concluding Remarks 93
References 94

6   Conclusions 97

Chapter 1
Introduction

There seems to be a strong demand for control schemes that include mechanical design criteria explicitly, e.g. in relation to active structural control during seismic excitation, [1–4]. Reduction of extreme response levels is the main objective, although the resulting decrease of response energy will also serve to prevent fatigue damage accumulation in structural components.

In the following, the excitation process under consideration is of the piecewise stationary type rather than being of a transient nature. The external loading and the structural response is further composed of a low-frequency component that is subjected to control and a high frequency component that is left unchanged.

The present focus is on interaction between structural reliability analysis and control algorithms. The main application area is positioning of floating vessels. The failure of the structural system is explicitly represented by the objective function and the derived control algorithms which are both of the on-line and pre-calibrated categories.

The intention of the present text is hence to give a certain background with respect to structural reliability analysis without going into the finer details. The main focus is on extreme response, but reliability assessment of the fatigue limit state is also addressed. Further description of structural reliability methods are given e.g. in [5, 6].

Furthermore, the background for classical optimal control algorithms is only given a rudimentary treatment in the present work. Further references to optimal control schemes are found e.g. in [7–9]. Optimal stochastic control as applied to structural systems is also addressed in [10–14].

The main objective of the following text is to bridge some of the gaps between mechanical engineering formulations and optimal stochastic control methods with focus on a particular area of application, which is floating vessels. It is intended that this outline may also be useful in connection with similar applications in other areas.

B. J. Leira, *Optimal Stochastic Control Schemes Within a Structural Reliability Framework*, SpringerBriefs in Statistics, DOI: 10.1007/978-3-319-01405-0_1, © The Author(s) 2013

References

1. Soong, T. T (1988). State-of-the-art review: Active control in civil engineering. Engineering Structures, *10*, 74–84.
2. Housner, G. W., Bergman, L. A., Caughey, T. K., Chassiakos, A. G., Claus, R. O., Masri, S. F., et al. (1997). Structural control: Past and present. *ASCE Journal of Engineering Mechanics, 123*, 897–971.
3. Spencer, B. F. Jr., Suhardjo, J., & Sain, M. K. (1994). Frequency domain optimal control strategies for aseismic protection. *Journal of Engineering Mechanics, 120*(1), 135–159.
4. Spencer, B. F. Jr., Timlin, T. L., & Sain, M. K. (1995). Series solution of a class of nonlinear optimal regulators. *Journal of Optimization Theory and Applications*.
5. Madsen, H., Krenk, S., & Lind, N. C. (1986). Methods of structural safety. New Jersey: Prentice-Hall, Englewood Cliffs.
6. Melchers, R. E. (1999). Structural reliability analysis and prediction. Chichester, UK: Wiley
7. Maybeck, P. S. (1982). Stochastic models, estimation and control, Vol 3, New York: Academic Press.
8. Fossen, T. I. (2002). *Marine control systems*. Trondheim, Norway: Marine Cybernetics.
9. Fuller, C. R., Elliott, S. J., & Nelson, P. A. (1997). Active control of vibration. Academic Press.
10. Casciati, F., Magonette, G., & Marazzi, F. (2006). Technology of semiactive devices and applications in vibration mitigation. Wiley.
11. Marti, K. (2008). Stochastic optimization problems, 2nd edition, Berlin-Heidelberg: Springer.
12. Preumont, A., & Seto, K. (2008). Active control of structures. Wiley.
13. Preumont, A. (2002). *Vibration control of active structures. An introduction* (2nd ed.). Dordrecht: Kluwer.
14. Wagg, D., & Neild, S. (2010). *Nonlinear vibration with control*. Dordrecht: Springer.

Chapter 2
Structural Limit States and Reliability Measures

Abstract In the present chapter the different levels of structural reliability methods are first summarized. Subsequently, the concept of the failure function is introduced and computation of the associated failure probability is considered. The interpretation of this probability as a volume is highlighted. Structures with time-varying load and resistance properties are next addressed. The simplified situation where both the random variable representing the load effect and the resistance are Gaussian is discussed. The resulting failure probability and the associated reliability index (beta-index) are elaborated.

Keywords Reliability methods · Failure function · Probability of failure · Reliability index

2.1 Introduction

Failure of a structure generally designates the event that the structure does not satisfy a specific set of functional requirements. Hence, it is a fairly wide concept which comprises such diversified phenomena as loss of stability, excessive response levels in terms of displacements, velocities or accelerations, as well as plastic deformations or fracture e.g. due to overload or fatigue.

The consequences of different types of failure also vary significantly. Collapse of a single sub-component does not necessarily imply that the structure as a system immediately loses the ability to carry the applied loads. At the other extreme, a sudden loss of stability is frequently accompanied by a complete and catastrophic collapse of the structure. Failure can also consist of a complex sequence of unfortunate events, possibly due to a juxtaposition of low-probability external or man-made actions and internal defects.

In engineering design, distinction is typically made between different categories of design criteria. These are frequently also referred to as *limit states*. The three most common categories are the Serviceability Limit State (SLS), the Ultimate

B. J. Leira, *Optimal Stochastic Control Schemes Within*
a Structural Reliability Framework, SpringerBriefs in Statistics,
DOI: 10.1007/978-3-319-01405-0_2, © The Author(s) 2013

Limit State (ULS) and the Fatigue Limit State (FLS). Many design documents also introduce the so-called Accidental Limit State and the Progressive Limit State in order to take care of unlikely but serious structural conditions.

Engineering design rules are generally classified as Level I reliability methods. These design procedures apply point values for the various design parameters and also introduce specific codified safety factors (also referred to as partial coefficients) which are intended to reflect the inherent statistical scatter associated in the parameters.

At the next level, second-order statistical information (i.e. information on variances and correlation properties in addition to mean values) can be applied if such is available. The resulting reliability measure and analysis method are then referred to as a Level II reliability method. At Level III, it is assumed that a complete set of probabilistic information (i.e. in the form of joint density and distribution functions) is at hand.

2.2 Failure Function and Probability of Failure

The common basis for the different levels of reliability methods is the introduction of a so-called failure function (or limit state function, or g-function) which gives a mathematical definition of the failure event in mechanical terms. In order to be able to estimate the failure probability, it is necessary to know the difference between the maximum load a structure is able to withstand, R (often referred to as *resistance*), the loads it will be exposed to, Q, and the associated *load effects* S. The latter are typically obtained by means of (more or less) conventional structural analysis methods. For this "generic" case, the "g-function" is then expressed as:

$$g(R, S) = R - S \qquad (2.1)$$

For positive values of this function (i.e. for R > S), the structure is in a safe state. Hence, the associated parameter region is referred to as the *safe domain*. For negative values (i.e. R < S), the structure is in a failed condition. The associated parameter region is accordingly referred to as the *failure domain*. The boundary between these two regions is the failure surface (i.e. R = S). The reason for application of these generalized terms is that the scalar quantities R and S in most cases are functions of a number of more basic design parameters. This implies that the simplistic two-dimensional formulation in reality involves a much larger number of such parameters corresponding to a reliability formulation of (typically) high dimension.

Here, a brief introduction is given to the basis for the Level III structural reliability methods which are required in the subsequent sections. Further details of these methods are found e.g. in [1, 2]. When concerned with waves, wind and dynamic structural response, it is common to assume that the statistical parameters are constant over a time period with a duration of (at least) 1 h. This is frequently referred to as a *short term statistical analysis*. A further assumption is typically that the stochastic dynamic excitation processes (i.e. the surface elevation or the wind turbulence velocity) are of the Gaussian type.

If the joint probability density function (or distribution function) of the strength and the load effect, i.e. f_{RS} (r,s) is known, the probability of failure can generally be expressed as

$$p_f = P(Z = R - S \leq 0) = \iint\limits_{R \leq S} f_{R,S}(r, s)\, dr\, ds \qquad (2.2)$$

where the integration is to be performed over the failure domain, i.e. the region where the strength is smaller than or equal to the load effect.

This is illustrated in Fig. 2.1a, where both the joint density function and the two marginal density functions $f_R(r)$ and $f_S(s)$ are shown (the latter are obtained by a one-dimensioanl integration of the joint density function with respect to each of the variables from minus to plus infinity). The joint density function can then be split into two separate pieces as shown in part (b) and (c) of the same figure. The failure probability can now be interpreted in a geometric sense as the volume of the joint density function which is located in the failure domain, i.e. the part of the plane to the right of the line $R = S$ (i.e. the region for which $S > R$). This corresponds to the slice of the volume of the joint pdf which is shown in Fig. 2.1c.

For the case of independent variables, the joint density function is just expressed as the product of the two marginal density functions. The resulting expression for the failure probability then becomes:

$$p_f = P(Z = R - S \leq 0) = \iint\limits_{R \leq S} f_R(r) \cdot f_S(s)\, dr\, ds \qquad (2.3)$$

where it is assumed that R and S are independent. By performing the integration with respect to the resistance variable, this can also be expressed as

$$p_f = P(Z = R - S \leq 0) = \int\limits_{-\infty}^{+\infty} F_R(s) f_S(s)\, ds \qquad (2.4)$$

where

$$F_R(s) = P(R \leq s) = \int\limits_{-\infty}^{s} f_R(r)\, dr \qquad (2.5)$$

This situation is illustrated in Fig. 2.2 where the two marginal density functions now are shown in the same plane. The interval which contributes most to the failure probability is where both of the density functions have non-vanishing values (i.e. in the range between 1 and 3.5 for this particular example).

(a)

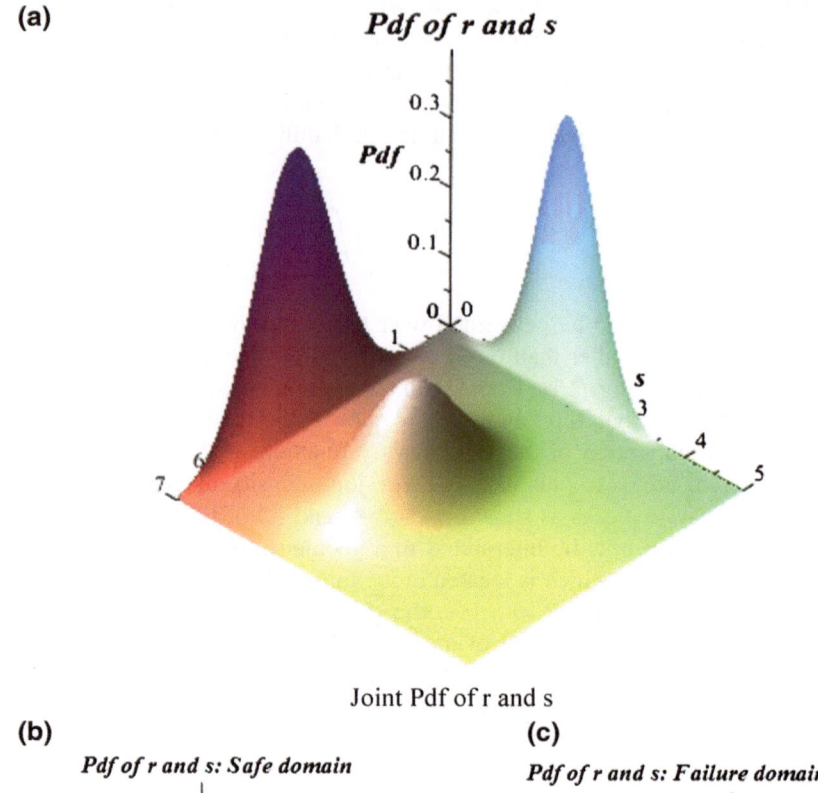

Joint Pdf of r and s

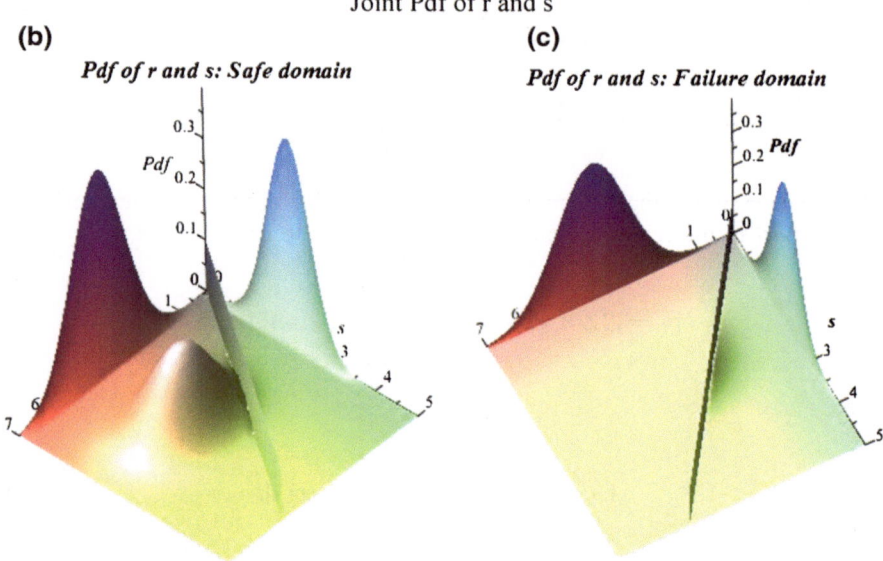

Safe domain volume   Failure domain volume

Fig. 2.1 Interpretation of failure probability as a volume. **a** Joint Pdf of r and s. **b** Safe domain volume. **c** Failure domain volume

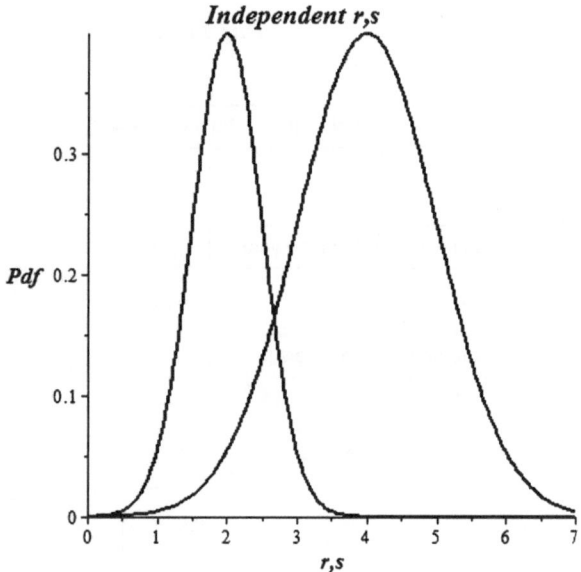

Fig. 2.2 Marginal densities projected into same plane for the case of independent variables

The integral in (2.4) is known as a convolution integral, where $F_R(r)$ denotes the cumulative distribution function of the mechanical resistance variable R. Closed-form expressions for this integral can be obtained for certain distributions, such as the Gauss distribution as will be discussed below. The resistance probability density function in Eq. (2.3), $f_R(r)$, is frequently represented as a Gaussian or Lognormal variable. The density function of the load-effect, $f_S(s)$, typically corresponds to extreme environmental conditions (such as wind and waves) and is frequently assumed to be described by a Gumbel distribution, see e.g. [3].

2.3 Time-Varying Load and Capacity

As the loads on marine structures are mainly due to wave-, wind and current, the statistical properties will fluctuate with time. The resistance will also in general be a function of time e.g. due to deterioration processes such as corrosion (this can clearly be counteracted by repair or other types of strength upgrading). Furthermore, a typical situation is that the extreme load effects increase with the duration of the time interval (i.e. the 20 year extreme value is higher than the 10 year extreme value, and the 3-h extreme load-effect during a storm is higher than the 1-h extreme load-effect).

This situation is illustrated by Fig. 2.3 for a relatively long time horizon. Here, t denotes time, and $t_0 = 0$ is the start time. The second "time slice" is at 10 years, and the third slice is at 20 years. The corresponding probability density functions

Pdf Time variation

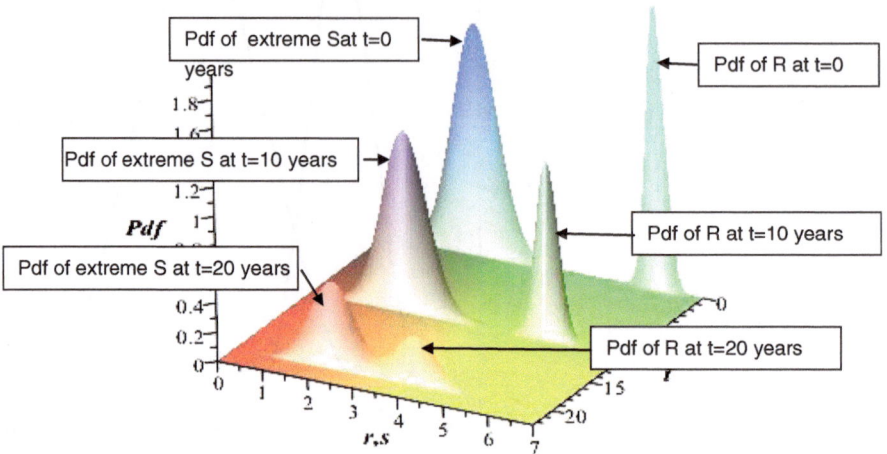

Fig. 2.3 Illustration of the time-varying marginal Pdfs of resistance, r(t), and extreme load-effect s(t)

of the mechanical resistance and the load effect are also shown for each of the three slices.

The figure illustrates that the structure will fail if (at any time during the considered time interval)

$$Z(t) = r(t) - s(t) < 0, \qquad (2.6)$$

where $Z(t)$ is referred to as the safety margin (which varies with time). The probability that the event described by (2.6) will take place can be evaluated from the amount of overlap by the two probability density functions $f_R(r)$ and $f_S(s)$ at each time step, as shown in Fig. 2.3. At $t = 0$ and 10 years, these two functions barely touch each other, while at $t = 20$ years, they have a significant amount of overlap. The latter case represents a corresponding increase of the failure probability.

If it is chosen to use time-independent values of either R or S (or both), the minimum value of (2.5) during the interval [0,T] must be used, where T denotes the design life time or the duration of a specific operation under consideration. In relation to the maximum load effect, an extreme value distribution, such as the Gumbel distribution, (also referred to as the type I asymptotic form) as mentioned above. The Gumbel distribution may be applied in cases where the initial distribution has an exponentially decaying tail which is the case e.g. for stochastic processes of the Gaussian type. Similarly, the probability density and distribution function of the minimum value is relevant. For durations of the order of a few days or less, simplifications can typically be introduced, since decrease of the strength properties on such limited time scales can usually be neglected.

The variation of the density functions will furthermore be different for the different types of limit states. For the fatigue limit state, the "resistance" can e.g. correspond to the permissible cumulative damage (i.e. given by a Miner-Palmgren sum equal to 1.0). This is a time-independent quantity which may still be represented by a (time-invariant) random variable. The "load-effect" will now correspond to the (random) cumulative damage which is obtained from the probability distribution of the stress range cycles. If there are other deterioration processes present (such as corrosion), the "resistance" will clearly decrease with time also for this type of limit state.

2.4 Simplified Calculation of Failure Probability for Gaussian Random Variables

As a special (and simplified) case, we next consider the situation when both R and S are Gaussian random variables, with mean values μ_R and μ_S and variances σ_R^2 and σ_S^2. Furthermore, the two variables are assumed to be uncorrelated. The quantity $Z = R-S$ will then also be Gaussian, with the mean value and variance being given by

$$\mu_Z = \mu_R - \mu_S$$
$$\text{and} \quad \sigma_Z^2 = \sigma_R^2 + \sigma_S^2 \tag{2.7}$$

The probability of failure may then be written as

$$p_f = P(Z = R - S \leq 0) = \Phi\left(\frac{0 - \mu_Z}{\sigma_Z}\right) = \Phi\left(-\frac{\mu_Z}{\sigma_Z}\right) \tag{2.8}$$

where $\Phi(.)$ is the standard normal distribution function (corresponding to a Gaussian variable with mean value 0 and standard deviation of 1.0). By inserting (2.6) into (2.7), we get

$$p_f = \Phi\left(-\frac{\{\mu_R - \mu_S\}}{\sqrt{\sigma_R^2 + \sigma_S^2}}\right) = \Phi(-\beta) \tag{2.9}$$

where

$$\beta = \frac{\{\mu_R - \mu_S\}}{\sqrt{\sigma_R^2 + \sigma_S^2}} \tag{2.10}$$

is defined as the safety index, see [4]. By defining an acceptable failure probability (i.e. $p_f = p_A$) on the left-hand side of this equation, one can find the corresponding value of β, i.e. β_A, that represents an acceptable lower bound on β (since decreasing β results in a higher failure probability). This value can be used to determine in a probabilistic sense whether the resistance R is within an acceptable range as compared to the load effect, S.

2.5 Calculation of Failure Probability for Non-Gaussian Random Variables

The index above can also be extended to handle reliability formulations which involve random vectors of arbitrary dimensions. Typically, both the load effect term, S, and the resistance term, R, are expressed as functions of a number of basic parameters of a random nature. Assembling these in the respective vectors $\mathbf{X_S}$ and $\mathbf{X_R}$, the failure function becomes a function of the vector $\mathbf{X} = [\mathbf{X_S^T}, \mathbf{X_R}]^T$. The failure surface will accordingly be defined by the equation $g(\mathbf{X}) = 0$.

The safety index can readily be extended to comprise correlated as well as non-gaussian variables. For general types of probability distributions, the failure probability as expressed by the integral in Eq. (2.1) can be computed e.g. by numerical integration. However, there also exist efficient approximate methods based on transformation into uncorrelated and standardized Gaussian variables.

In the case of non-gaussian and <u>uncorrelated</u> variables this transformation is based on the following expressions:

$$\Phi(u_1(t)) = F_{X_1}(x_1(t))$$
$$\vdots$$
$$\vdots \tag{2.11}$$
$$\Phi(u_n(t)) = F_{X_n}(x_n(t))$$

The simplified $g(R,S) = (R-S)$ reliability formulation may serve to illustrate how this transformation works for two different cases. As a first reference case, the two basic variables are taken to be uncorrelated Gaussian variables with mean values $(\mu_R = 3.0, \mu_S = 1.0)$ and standard deviations $(\sigma_R = 0.1, \sigma_S = 0.2)$. The relationship between the original basic variables and the transformed standardized Gaussian variables are then simply expressed as $R = 0.1\ U_1 + 3$ and $S = 0.2\ U_2 + 1$. The corresponding failure function in the transformed and normalized (U_1, U_2)-plane is shown as the upper surface in Fig. 2.4.

As a second case, the basic random variables R and S are both taken to be uncorrelated lognormal variables with the same mean values and standard deviations as before. The two corresponding transformations based on Eq. (2.11) then are expressed as:

$$\ln(r) = \sigma_{z1}u_1 + \mu_{z1} \quad \ln(s) = \sigma_{z2}u_2 + \mu_{z2}$$

where

$$\left(\sigma_{z1}^2 = \ln\left(1 + \left(\frac{\sigma_R^2}{\mu_R}\right)\right) = 0.0011, \sigma_{z2}^2 = \ln\left(1 + \left(\frac{\sigma_S^2}{\mu_S}\right)\right) = 0.039\right)$$

Fig. 2.4 Comparison of failure functions expressed in terms of transformed (normalized) Gaussian variables u1 and u2. *Upper* Gaussian basic random variables. *Lower* Lognormal basic random variables

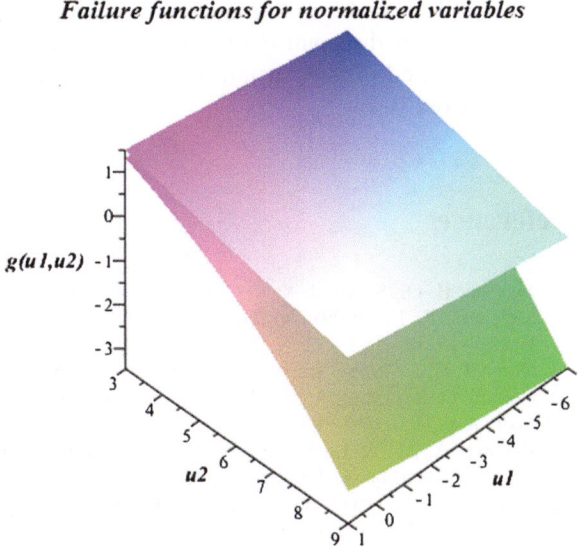

Failure functions for normalized variables

and

$$\left(\mu_{z1} = \ln(\mu_R) - 0.5\sigma_{z1}^2 = 1.098, \mu_{z2} = \ln(\mu_S) - 0.5\sigma_{z2}^2 = -0.0196\right)$$

The corresponding failure function $g(u_1,u_2) = \exp(\sigma_{z1}u_1 + \mu_{Z1}) - \exp(\sigma_{z2}u_2 + \mu_{Z2})$ is shown in the right part of Fig. 2.4.

Transformation of <u>correlated</u> non-gaussian variables requires that both marginal and conditional distribution functions are applied. For a number of n random variables the expressions become:

$$\Phi(u_1(t)) = F_{X_1}(x_1(t))$$
$$..$$
$$..$$ \hfill (2.12)
$$\Phi(u_n(t)) = F_{X_n|X_1,X_2,...,X_{n-1}}\left(x_n(t)|x_1(t), x_2(t), ...x_{(n-1)}(t)\right)$$

where the conditional cumulative distribution functions of increasing order are required.

Computation of the failure probability is frequently based on approximating the failure surface by its tangent plane at a proper point, or a second order surface at the same point. This point is identified by means of numerical iteration and is the point which is closest to the origin in the transformed space of standardized Gaussian variables (i.e. all of which have mean value zero and unit standard deviation).

For a more detailed description of these procedures (i.e. related to transformation of the variables and searching for the design point), reference is e.g. made to [1, 2, 5–12].

It is highly relevant to introduce a simplified version of the reliability index in Eq. (2.9) within the context of application within an on-line control algorithm. This is discussed in more detail in relation to the examples of application which are considered below.

References

1. Madsen, H. O., S. Krenk & N. C. Lind (1986): *Methods of Structural Safety*, Prentice-Hall.
2. Melchers, R. E. (1999). *Structural reliability: Analysis and prediction*. Chichester: Ellis-Horwood Ltd.
3. Gumbel, E. J. (1958). *Statistics of extremes*. New York, US: Columbia University Press.
4. Cornell, C. A. (1969). A probability-based structural code, *Journal of the American Concrete Institute 60*(12), 974–985.
5. Hasofer, A. M., & Lind, N. C. (1974). Exact and invariant second moment code format, ASCE. *Journal of the Engineering Mechanics Division, 100*, 111–121.
6. Ditlevsen, O. (1981). Principle of normal tail approximation, ASCE. *Journal of the Engineering Mechanics Division, 107*, 1191–1208.
7. Hohenbichler, M., & Rackwitz, R. (1981). Non-normal dependent vectors in structural safety. ASCE. *Journal of the Engineering Mechanics Division, 107*, 1227–1258.
8. Hohenbichler, M., & Rackwitz, R. (1983). First-order concepts in system reliability. *Structural Safety, 1*, 177–188.
9. Rosenblatt, M. (1952). Remarks on a Multivariate Transformation. *The Annals of Mathematical Statistics, 23*, 470–472.
10. Breitung, K. (1984). Asymptotic approximations for multinormal integrals, ASCE. *Journal of the Engineering Mechanics Division, 110*, 357–366.
11. Tvedt, L. (1989). Second order reliability by an exact integral. *Lecture Notes in Engineering, 48*, Springer, pp. 377–384.
12. Der Kiureghian, A., & Liu, P. L. (1986). Structural reliability under incomplete probability information, ASCE. *Journal of the Engineering Mechanics Division, 112*(1), 85–104.

Chapter 3
Dynamic Structural Response Analysis and Probabilistic Representation

Abstract The present chapter summarizes probabilistic modeling related to environmental processes and associated stochastic dynamic structural response. Distinction is made between so-called "short-term" and "long-term" analysis. The former refers to time intervals with a duration of the order of a few hours for which stationary conditions can be assumed. The latter corresponds to the basic variation of the environmental processes on more macroscopic time scales. Methods for dynamic response analysis are briefly summarized, and associated structural reliability formulations associated with both scalar and multi-component stochastic response processes are highlighted.

Keywords Environmental processes · Stochastic dynamic response · Time-invariant reliability

3.1 General

As observed in Chap. 2, quantification of structural reliability requires that statistical properties of both load effects and structural capacity are available. In the present Chapter, we focus on the load effects since this will be most relevant within a control-oriented framework. The random variables associated with the capacity term (or resistance term), R, in the previous chapter is not treated in any detail in the following. However, further information can be found e.g., in [1, 2].

For structures which are subjected to time-varying loading, the load effect will also be time-dependent, i.e., $S = S(t)$. Within the present context, the load is further of a stochastic nature which implies that an ensemble of load effect histories (response realizations or sample functions) are relevant. Typically a set of selected characteristic parameters related to the response are relevant rather than the complete time history itself. Examples of such "characteristic parameters" are extreme response levels during a specific time period, or stress cycle distributions which are relevant for computation of accumulated fatigue damage.

B. J. Leira, *Optimal Stochastic Control Schemes within a Structural Reliability Framework*, SpringerBriefs in Statistics, DOI: 10.1007/978-3-319-01405-0_3, © The Author(s) 2013

Such time-varying loading can be caused by environmental processes of different types, such as earthquakes, waves, wind, current-induced vortices, traffic loading or machine-induced vibrations. Focus is presently on the marine environment which implies that wind, waves and current are the most relevant sources of dynamic loading.

The structural response to time-varying loading will sometimes be of a static or quasi-static nature. However, in general dynamic effects will be of importance and accordingly need to be taken into account.

In the following, we first review briefly statistical models of the environment and subsequently turn our attention to statistical representation of the response. For both cases, scalar as well as multidimensional models are highlighted.

3.2 Environmental Modeling

Due to the non-stationary nature of most environmental processes, the corresponding modeling typically consists of two main building blocks which correspond to the so-called "short-term" and "long-term" behavior. The short-term modeling is associated with "stationary" conditions which are conditional on given values of particular characteristic environmental parameters.

Models related to the "long-term" representation of environmental processes will typically involve joint statistical modeling of several characteristic parameters. As an example, the wave climate is typically represented by the significant wave height and a characteristic period (e.g., peak period or zero-crossing period). The wind climate is similarly characterized by the mean wind and turbulence intensity. (The expression "long-term" can in the present context also refer to a subset of environmental conditions, and can hence reflect e.g., seasonal environmental characteristics which are relevant for marine operations).

Knowledge of the joint statistical properties of two or several simultaneous environmental parameters will accordingly play an important role for many activities at sea. This applies both for the open ocean and coastal areas.

In particular, the bivariate probability distribution of significant wave height and characteristic period is highly relevant for a number of applications, see e.g., [3, 4]. The significant wave height characterizes the intensity of the sea states, while the mean period or the peak period is relevant for assessing the possibility of exciting the natural periods of a given structure. Hence, the joint distribution of significant wave height and characteristic period is required in order to address several issues which are relevant for design of marine and offshore structures. Furthermore, it represents a key issue in connection with planning of marine operations.

An example of a joint pdf of significant wave height (H_s) and peak period (T_p) based on a data set given in [4] is shown in Fig. 3.1. The correlation between the two parameters is clearly reflected in the shape of the pdf and its associated level curves.

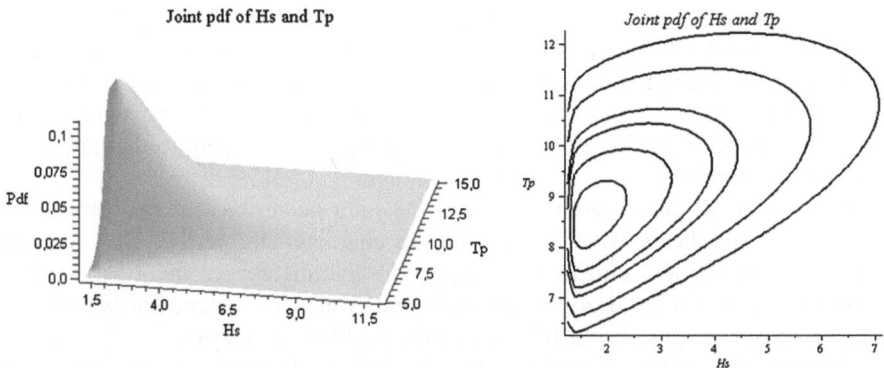

Fig. 3.1 Example of joint pdf of wave height and peak period based on data given in Bitner-Gregersen and Guedes Soares [4] (Pdf contour levels are at [0.01, 0.02, 0.04, 0.05, 0.07, 0.095])

A number of other statistical models have also been applied in order to model the joint behavior of these variables. Ochi [5], has adopted a bivariate Lognormal distribution, which implies an exponential transformation of the bivariate normal distribution. This model has the great advantage of simplicity but is not always quite accurate in the upper "low probability" range.

To ensure a good fit to the data Haver [6], has chosen separate models for the significant wave height (H_s) and the peak period (T_p). A combination of a Weibull and a Lognormal distribution was applied in order to model the marginal distribution of significant wave height. The marginal distribution of the peak period was fitted by a Lognormal distribution. A regression equation was proposed for the parameters of the conditional distributions of the peak period as a function of significant wave height in order to be able to extrapolate the parameters at the low probability end.

Mathiesen and Bitner-Gregersen [7], applied a three-parameter Weibull distribution to model the marginal distribution of significant wave height. This was combined with the conditional Lognormal distribution for wave periods. They compared this model with the bivariate Lognormal and with a bivariate Weibull distribution. It was concluded that the approach with the conditional distribution of wave period provided the best fit to data.

Athanassoulis et al. [8], have proposed an approach that combines some degree of flexibility with a certain simplicity and parsimony in the number of required parameters. They use the Plackett bivariate model to fit bivariate distribution functions to the data. This model, even though not being completely general, allows the specification of any two marginal distributions and accounts for the dependence structure by means of a parameter related to the correlation between the variables.

Ferreira and Guedes Soares [9], decided to use the bivariate normal distribution to model the data after transformation and they have chosen the Box-Cox transformation to make the variables close to normally distributed. This transformation has been applied in several univariate and bivariate autoregressive models of wave time series. Prince-Wright [10], applied an extended version of this transformation to the data. In Soares and Guedes Soares [11], it was also concluded that the Box-Cox model may be a good choice for many applications, as it represents a compromise between accuracy of the fit and simplicity with respect to the number of parameters involved.

The models discussed above apply to the joint occurrence of two parameters such as the significant wave height and the characteristic period. However, the literature contains less information when more parameters are involved. Most of these models use a conditional approach that results in a rapid increase in the amount of data required as a function of the problem dimension.

Other studies on joint modeling have also been performed, see e.g., Nerzic and Prevosto [12], in connection with the Plackett model or Prince-Wright [10], in relation to the Johnson transformations (see e.g., Johnson and Kotz [13]). In Fouques et al. [14], several metocean parameters are represented in the model such as the significant wave height, the mean wave period, the mean wind velocity and sea state persistence parameters. Seasonal models for the different parameters were also introduced.

Multivariate extreme-value models for significant wave height and peak period have been considered by Jonathan et al. [15].

3.3 Dynamic Response Analysis

The main characteristics of the most common methods for structural response analysis are basically the same both for deterministic and stochastic load models. However, the practical implementation of relevant analysis procedures may differ somewhat. This applies in particular to the evaluation of probabilistic response properties in the case of stochastic loading combined with non-linear structural behavior.

Two main categories of methods for dynamic response analysis are provided by the *time and frequency domain* approaches. The former is most relevant in order to study the effect of non-linear effects associated with loads and structural behavior. Time domain analysis in connection with stochastic loading is generally based on generating a number of sample functions of this loading and subsequently computing the associated sample time histories for the response quantities of interest.

In connection with assessment of control schemes, the time domain approach is much applied. Structural response analysis in the time-domain is based on step-wise integration of the dynamic equilibrium equation, see e.g., Clough and Penzien [16], Newland [17]. Further details of the different terms of this equation associated with floating structures are given e.g., in Fossen [18]. In the case of non-linear structural behaviour, the incremental form of this equilibrium equation is quite commonly applied. Classical references in relation to assessment of numerical stability related to step-wise time integration in structural dynamic analyses are e.g., Newmark [19], Belytscho and Shoeberle [20], and Hughes [21, 22]. More recent studies are summarized e.g., in Bathe [23]. Elaboration of suitable time integration methods for systems with large displacements and constraint conditions can be found e.g. in Krenk [24].

The frequency domain approach is most relevant in connection with linear (or linearized) models of load and structural behavior, and it is generally superior in terms of computation time as well as manual processing time. However, this type of response analysis is not so relevant in connection with on-line control schemes. Further details are found e.g., in Newland [17].

For evaluation of the so-called quasistatic response, dynamic effects are neglected. This implies that the response is obtained by inverting the stiffness matrix based on a numerical model of the structure. This type of analysis can e.g., be relevant if the frequencies associated with the loading are much lower than the natural frequencies of the structure.

For structural response analysis in connection with *stochastic* loading additional tools are also available for a more direct evaluation of specific probabilistic features of the response. Examples of such "probability domain" methods are the so-called covariance analysis and moment equations. A complete solution method in terms of probability density functions c an be achieved by solution of the Fokker–Planck equation. For a classical reference related to the latter approach, see e.g., Risken [25]. A quite comprehensive overview of different types of solution methods is provided by Kumar and Narayanan [26]. Challenges related to higher-dimensional formulations of this equation is discussed by Masud and Bergman [27].

As an introduction to the subject of response statistics in relation to stochastic processes, focus is on scalar processes of the Gaussian type. An overview of relevant statistical distributions for such processes are given in the following section.

3.4 Response Statistics for Scalar Gaussian Processes

Similar to the environmental processes, analysis of structural response processes are generally also based on a distinction between "short-term" and "long-term" models. In the following, we first review relevant statistical properties related to the "short-term" modeling, i.e., for stationary conditions.

In connection with formulation of failure functions for structural reliability, the probability distributions of local response maxima and extreme values are highly relevant. Furthermore, for the fatigue limit state the probability distribution of stress cycles is required. This distribution is closely related to the probability distribution of local maxima. These topics are addressed in the following.

The probability distribution of local maxima for a scalar Gaussian process was derived by Rice [28], Longuet-Higgins [29], and Cartwright and Longuet-Higgins [30], and is referred to as the Rice distribution. The shape of this distribution depends strongly on the so-called bandwidth parameter, ε, which is defined as

$$\varepsilon = \sqrt{1 - \frac{\dot{\sigma}_x^2}{\sigma_x^2 \ddot{\sigma}_x^2}} \qquad (3.1)$$

where σ_x^2 is the variance of the response process, $\dot{\sigma}_x^2$ is the variance of the associated velocity process and $\ddot{\sigma}_x^2$ is the variance of the response acceleration process.

For a so-called wide-band process (for which the bandwidth parameter approaches 1.0), the Rice distribution approaches the Gaussian distribution in the limit. The associated mean value is zero, which implies that positive and negative local maxima are equally likely. For a so-called narrow-band process (for which the band-width parameter approaches 0.), instead the Rayleigh distribution applies for the local maxima. The corresponding distribution function is given by

$$F_S(s) = 1 - \exp\left(-\frac{s^2}{2\sigma_x^2}\right) \qquad (3.2)$$

where s is the magnitude of the local maximum. The associated density function becomes

$$f_S(s) = \frac{s}{\sigma_x^2} \exp\left(-\frac{s^2}{2\sigma_x^2}\right) \qquad (3.3)$$

The distribution of local maxima can also be obtained by means of the so-called Powell approximation by utilization of the so-called up-crossing rate for the level s, which is denoted by $v_x^+(s)$. The so-called zero-crossing rate is obtained by setting $s = 0$, i.e., $v_x^+(0)$. The distribution of local maxima based on the Powell approximation is then expressed as:

$$F_{S,\text{Powell}}(s) = 1 - \left(\frac{v_x^+(s)}{v_{x,\max}^+}\right) \qquad (3.4)$$

where $v_{x,\max}^+$ is the maximum possible value of the up-crossing rate.

For a Gaussian process the up-crossing rate is expressed as

$$v_x^+(s) = \frac{\dot{\sigma}_x}{2\pi\sigma_x} \exp\left(-\frac{s^2}{2\sigma_x^2}\right) \qquad (3.5)$$

and the maximum value occurs just for the level $s = 0$. Accordingly, the resulting cumulative distribution of local maxima based on the Powell approximation is expressed as

$$F_{S,\text{Powell}}(s) = 1 - \left(\frac{v_x^+(s)}{v_{x,\max}^+}\right) = 1 - \exp\left(-\frac{s^2}{2\sigma_x^2}\right) \qquad (3.6)$$

which is identical with the Rayleigh distribution, i.e., the narrow-band limit of the Rice distribution.

Based on the distribution function for local maxima, the corresponding extreme value distribution for a given duration T can also be obtained. The number of local maxima for a narrow-band process during this period can be estimated based on the zero-crossing frequency as: $N = v_x^+(0)T = \frac{\dot{\sigma}_x T}{2\pi\sigma_x}$. By further assuming that the local maxima are statistically independent, the cumulative distribution function of the extreme value during T (i.e., which is referred to as $X_{E,T}$) is obtained as

$$F_{X_{E,T}}(x_{E,T}) = (F_s(x_{E,T}))^N = \left(1 - \left(\frac{v_x^+(x_{E,T})}{v_x^+(0)}\right)\right)^N = \left(1 - \exp\left(-\frac{x_{E,T}^2}{2\sigma_x^2}\right)\right)^N$$

(3.7)

The corresponding density function is readily obtained by differentiation. A plot of this density function is shown in Fig. 3.2 below for increasing values of the exponent N (in the range from 50 to 5,000). The x-axis in the figure corresponds to the normalized variable, i.e., $z = \frac{x_{E,T}}{\sigma_x}$. The figure clearly shows the increase of the mean value for increasing values of the exponent N.

We also note that when N approaches infinity, the distribution function in Eq. (3.7) can be rewritten as:

$$\lim_{N\to\infty}(F_{X_{E,T}}(x_{E,T})) = \lim_{N\to\infty}(F_s(x_{E,T}))^N = \lim_{N\to\infty}\left(1 - \left(\frac{Tv_x^+(x_{E,T})}{Tv_x^+(0)}\right)\right)^N$$

$$= \lim_{N\to\infty}\left(1 - \left(\frac{Tv_x^+(x_{E,T})}{N}\right)\right)^N$$

(3.8)

$$= \exp(-Tv_x^+(x_{E,T})) = \exp\left(-\left(\frac{\dot{\sigma}_x T}{2\pi\sigma_x}\right)\exp\left(-\frac{x_{E,T}^2}{2\sigma_x^2}\right)\right)$$

For high levels, this expression can subsequently be approximated by a Gumbel distribution function of the following form:

$$F_{X_{E,T}}(x_{E,T}) = \exp-(\exp(-\alpha(x_{E,T} - u)))$$

(3.9)

where α and u are the two parameters (i.e., constants) of the Gumbel distribution function. The proper expressions for these parameters are now obtained by setting the second exponential term for each of the two distribution functions equal to each other, i.e.,

$$(\exp(-\alpha(x_{E,T} - u))) = \left(\left(\frac{\dot{\sigma}_x T}{2\pi\sigma_x}\right)\exp\left(-\frac{x_{E,T}^2}{2\sigma_x^2}\right)\right) = \left(N\exp\left(-\frac{x_{E,T}^2}{2\sigma_x^2}\right)\right)$$

(3.10)

By taking the logarithm of both sides of this equation we obtain:

$$(-\alpha(x_{E,T} - u)) \approx (\ln(N)) - \frac{x_{E,T}^2}{2\sigma_x^2}$$

(3.11)

Extreme pdf vs. z and N

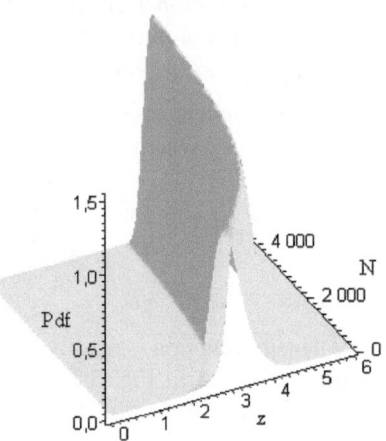

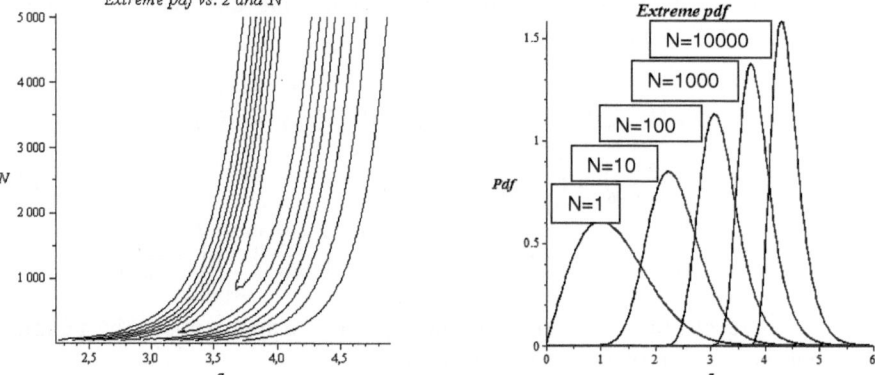

Fig. 3.2 Extreme value density function for increasing N (i.e., number of local maxima for the parent Gaussian process)

The Gumbel density function has a peak at $x_{E,T} = u$, and hence we can also perform a Taylor series expansion of the quadratic term around this value:

$$\frac{x_{E,T}^2}{2\sigma_x^2} \approx \frac{1}{2\sigma_x^2}\left(u^2 + 2u(x - u)\right) = \frac{1}{2\sigma_x^2}\left(2ux - u^2\right) \tag{3.12}$$

By inserting this expression in Eq. (3.11) above, this gives two equations for the two constants α and u. These equation are solved to give:

$$\alpha = \frac{u}{\sigma_x^2} \quad \text{and} \quad \alpha u = \left(\frac{u}{\sigma_x^2}\right)u = \left(\frac{u^2}{\sigma_x^2}\right) = \ln(N) + \frac{1}{2}\left(\frac{u^2}{\sigma_x^2}\right) \tag{3.13}$$

The last equality in the second of these equations then gives $u = \sigma_x\sqrt{2\ln(N)}$.

This value is just equal to the so-called *characteristic (or most likely) largest* value. The corresponding *expected largest value* for the Gumbel distribution is obtained from the expression that applies to that particular distribution by inserting the above values for u and α:

$$E[x_{E,T}] = u + \frac{0.5772}{\alpha} = \sigma_x\left(\sqrt{2\ln(N)} + \frac{0.5772}{\sqrt{2\ln(N)}}\right) \tag{3.14}$$

where $\gamma = 0.5772$ is the Euler constant.

The long-term distribution of response amplitudes is obtained by considering that for a random point in time the sea state parameters (contained in a vector $\mathbf{X}$) are themselves random variables with joint probability density function $f_x(x)$. The relevant types of joint density function for significant wave height and peak period were discussed above. The corresponding probability distribution of the response (i.e., *r*) is obtained by weighting the *conditional* distribution of local maxima by multiplying with the joint density function of sea state parameters and subsequently integrating:

$$F_{R,L}(r) = \int\limits_x F_S(r|\mathbf{x})f_\mathbf{x}(\mathbf{x})w(\mathbf{x})d\mathbf{x} \tag{3.15}$$

Here, $w(\mathbf{x})$ is a weighting factor which accounts for the relative number of response peaks in each sea state, and $F_S(r|\mathbf{x})$ is the short-term conditional response distribution for a specific sea state corresponding to a given outcome of the vector $\mathbf{X}$. The conditional distribution of local maxima is typically taken to be of the Rice, Rayleigh or Weibull type.

A similar expression also holds for the complement of the cumulative long-term distribution. This complementary distribution corresponds to the probability of exceeding a specific response levels. This probability is obtained by replacing the short-term distributions of the response amplitudes by their complements in Eq. (3.15).

A very convenient and generally applied approximation to the long-term distribution $F_{R,L}(r)$ is provided by the Weibull model.

Estimation of extreme values (e.g., 10 and 100 year wave amplitudes and wave heights) can be performed based on the long-term distribution by application of the proper probability level.

Alternatively, extreme response levels can also be estimated based on the corresponding extreme value distributions for each of the "short-term" conditions (i.e., for all the different sea states). The short- and long-term distributions in Eq. (3.15) are then replaced by the corresponding short- and long-term extreme-value distributions, i.e., $F_{S,E}(r_E|\mathbf{x})$ and $F_{L,E}(r_E)$, where the subscript E refers to extreme value:

$$F_{L,E}(r_E) = \int\limits_x F_{S,E}(r_E|\mathbf{x})f_\mathbf{x}(\mathbf{x})d\mathbf{x} \tag{3.16}$$

The resulting extreme response which results by application of this expression will typically agree quite well with that obtained by application of the long-term distribution.

A more correct formulation of the long-term extreme-value distribution based on direct application of the upcrossing-rate can be expressed as:

$$F_{L,E}(r_E) = \exp\left\{-\left(T \int_{x} v^+(r_E|\mathbf{x})f_{\mathbf{x}}(\mathbf{x})d\mathbf{x}\right)\right\} \tag{3.17}$$

It is anticipated that for most cases the difference between results obtained by application of Eq. (3.17) versus Eqs. (3.15) and (3.16) is negligible. Methods for computation of the upcrossing rate for Gaussian as well as non-Gaussian processes are found e.g., in Wen and Chen [31], Hagen and Tvedt [32], and Beck and Melchers [33].

Analysis of accumulated fatigue damage in a structural component based on the SN-curve approach is generally based on the following equation for the *expected* damage:

$$E[D(T)] = \frac{N(T)}{\bar{a}}E[(\Delta\sigma)^m] = \frac{N(T)}{\bar{a}} \int_0^\infty (\Delta\sigma)^m f_{\Delta\sigma}(\Delta\sigma)d(\Delta\sigma) \tag{3.18}$$

where the probability density function of the stress-range, i.e., $f_{\Delta\sigma}(\Delta\sigma)$, may apply to one specific sea state under consideration or may correspond to the long-term cycle distributions. In the former case, summation across the range of possible sea states is required. The probability density function for the stress range is frequently taken to be of the Weibull type, both within a short-term and long-term framework. The quantities $(\bar{a}, m)$ are constants which define a particular SN-curve, and N(T) is the expected number of stress cycles (or local maxima) during the time period T. For S–N curves with one and two slopes, analytical formulas for the resulting fatigue damage are found e.g., in Almar-Naess et al. [34].

3.5 Reliability Formulations for Scalar Gaussian Processes

In order to connect the topic of response statistics to the structural reliability analysis formulated in Chap. 2, the resistance (i.e., the response threshold) must in general also be represented as a random variable.

Reliability analysis in relation to the extreme response level exceeding the capacity threshold can then be based on the failure function from Eq. (2.6) in Chap. 2:

$$g(R, S) = R - S(t) \tag{3.19}$$

where the capacity threshold presently is represented by a time-invariant random variable. The time varying load term S(t) can also be replaced by its extreme value (i.e., $S_{E,T}$) during a specific time interval, T, by means of the formulation presented in Sect. 3.4. For a short-term condition this gives:

$$g(R, S) = R - S_{E,T} \tag{3.20}$$

where the load effect distribution function is given by Eq. (3.9) above. For a long-term analysis the corresponding distribution function is given by Eqs. (3.16) and (3.17).

For reliability formulations involving a deteriorating threshold, a sequence of stationary conditions need to be analyzed. A conservative approximation which leads to simplified calculations can be based on application of the statistical resistance parameters which correspond to the end of the time interval under consideration. This implies that the extreme loading is assumed to occur at the time when the capacity reaches its lowest value. An outline of different formulations which are relevant is given e.g., in Schall et. al. [35].

For reliability analysis related to the fatigue failure mode, the load effect term in Eq. (3.19) is instead expressed by Eq. (3.18). Due to the parameters which are input to the fatigue damage calculation being random variables, the left-hand side of the expression is also a random variable. The permissible threshold value is 1.0 for the fatigue damage (i.e., corresponding to R in Eq. 3.19) which is based on the Miner-Palmgren hypothesis for summation of partial damage contributions. However, as there is a significant model uncertainty related to this summation method this threshold value is also generally represented as a random variable (with a mean value of 1.0).

Generic models are frequently based on lognormal probability models for both the resistance and load terms. Based on such a model the resulting failure function can be expressed as:

$$g(R, S) = R - S(t) = R - D(t) \tag{3.21}$$

where R represents the lognormal distributed capacity with a mean value of 1.0, and the lognormal variable D(t) corresponds to the accumulated fatigue damage. The statistical parameters of the latter are generally functions of time. The resulting density functions and their relative location as a function of time are illustrated in Fig. 3.3 below. The mean value of the damage increases linearly with time (i.e., $\mu_D(t) = 0.004 \cdot t$ for the present example), and the variance is taken to increase in proportion to $t^{1.5}$ (i.e., $\sigma_D^2(t) = 0.00002 \cdot t^{1.5}$ for the present example): It is seen that the probability density function of the accumulated damage overlaps increasingly with the density function of the permissible damage. At time unit $t = 100$ the probability of failure calculated according to Eq. (2.2) is computed as $p_f = 0.02$ which reflects the proximity of the density functions at that time.

It is also possible to express the fatigue damage as a function of more basic input parameters, e.g., by means of response surface techniques. For a specific example of application of such a procedure for fatigue reliability analysis of

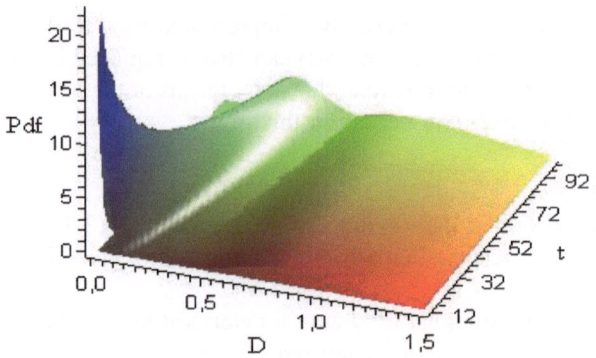

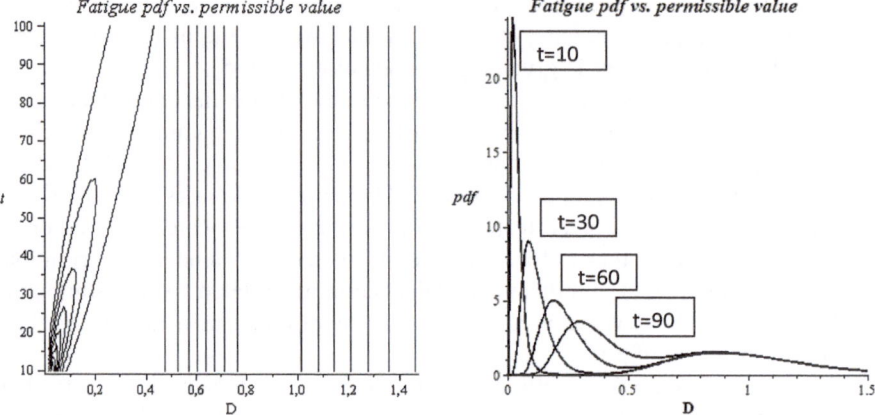

Fig. 3.3 Example of time variation of probability density functions related to fatigue reliability

marine risers based on response surface techniques, see e.g., Leira et al. [36]. A main feature of such structural components is significant dynamic response behavior.

3.6 Multi-Dimensional Response Statistics

3.6.1 Introduction

The statistical analysis of response processes with multiple components (i.e., vector-processes) is much less developed than for scalar processes. This applies in particular to non-Gaussian processes. The most consistent representation related to reliability analysis associated with such multi-dimensional processes is provided by the so-called outcrossing-rate. This quantity is first considered, and subsequently, some particular multivariate statistical formulations are summarized.

3.6.2 Outcrossing-Rate Analysis

The up-crossing rate which was introduced for scalar stochastic processes has been generalized to the so-called out-crossing rate which applies to vector-processes. Computation of this quantity requires that the joint probability density of the component processes and their velocities are known (i.e., for each stationary "short-term" condition). Subsequently, the joint density of the component processes and the *normal* velocity at a given point of the failure surface, $f_{\mathbf{X}, \dot{X}_n}(\mathbf{X}, \dot{X}_n)$, can be obtained. The outcrossing-rate from a surface ∂B which bounds the safe domain is then expressed by the generalized Rice's formula as (see e.g., Veneziano et al. [37]):

$$v^+(\partial B) = \int_{\partial B} \left\{ \int_0^\infty \dot{x} n f_{x \dot{x}_n}(x, x_n) d\dot{x}_n \right\} ds_1 ds_2 \ldots ds_{n-1} \qquad (3.22)$$

where the $(n-1)$ dimensional integration is performed over the local coordinates for the surface ∂B in n-dimensional space. If X and $\dot{X}_n$ are statistically uncorrelated processes this integral can be decomposed into a product of two simpler factors.

The out-crossing rate for a two-component Gaussian process for a straight line of infinite extension is shown in Fig. 3.4 as a function of the coordinates of the projection point (i.e., the projection from the origin to the straight line). In the left part of the figure, both of the two components have the same statistical properties. Furthermore, they are both normalized and uncorrelated. In the right part of the figure, the same properties apply except that the velocity variance for the X_2-

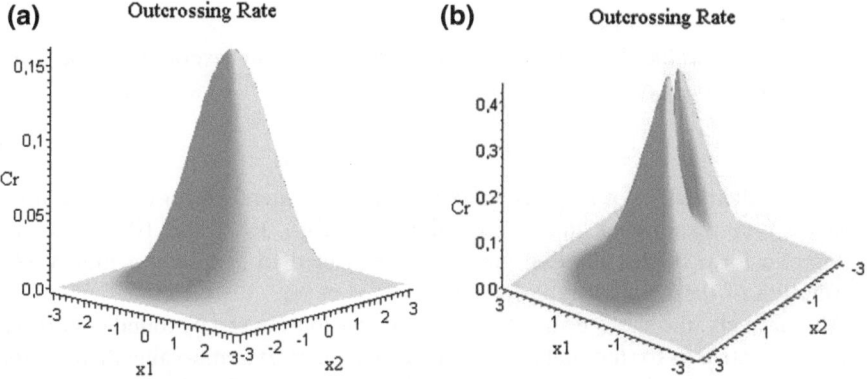

Fig. 3.4 Outcrossing-rate for straight line of infinite extension as function of projection point coordinates. **a** Uniform components, **b** Velocity-variance ratio is 8

component is 8 times that of the X_1-component. In the former case there is a uniformly rounded shape, while for the latter case a two-lobed shape (which is symmetric about the origin) is observed.

3.6.3 Local Maxima and Extremes

The analysis of multivariate stochastic processes can be much simplified if the ratios of the respective spectral moments of increasing order are identical for all the response process components which are involved. This implies that the associated characteristic periods (e.g., zero-crossing periods) will be identical. Furthermore, the number of local maxima which occur within a given time interval will be the same for all the components.

Such response processes are herein referred to as "iso-chromatic" with reference to the feature that the frequency distribution of the response energy will be similar for all the components. Hence, the "colour" of all the response components is also constant as indicated by the present notation.

For such cases, multivariate generalizations of the Rice, Rayleigh and Weibull distributions can be applied for representation of the joint local maxima and the cycle distributions.

If we consider a vector of iso-chromatic response processes, $\mathbf{Y}$, the marginal distribution of local maxima for each component can be expressed by the one-dimensional Rice, Rayleigh or Weibull distribution. Based on the correlation matrix, the conditional joint short-term distribution function of the same type (i.e., for given values of the wave parameters), $F_{\mathbf{Y}|\mathbf{X}}(\mathbf{y}|\mathbf{x})$, can also be established. The long term distribution of the local response maxima can then be expressed on a similar form as for the scalar response process above:

$$F_L(\mathbf{y}) = \int_{\mathbf{x}} F_{\mathbf{Y}|\mathbf{X}}(\mathbf{y}|\mathbf{x})f_{\mathbf{x}}(\mathbf{x})w(\mathbf{x})d\mathbf{x} \qquad (3.23)$$

The extreme-value distribution based on the long-term response can be found by exponentiation, i.e.,

$$F_E(\mathbf{y}) = [F_L(\mathbf{y})]^N \qquad (3.24)$$

where N is the number of response maxima (which is identical for all the response processes) corresponding to a given duration, e.g., 1, 10 or 100 years. The asymptotic form of this distribution can be expected to approach a multivariate distribution of the Gumbel type for large values of N.

Alternatively, the short term extreme-value distribution can be found conditional on each sea-state as for the scalar case. Such short-term extreme-value distributions have been considered e.g., by Gupta and Manohar [38]. The long-term extreme value distribution is subsequently computed by integration across the "long-term" environmental characteristics:

$$F_{L,E}(\mathbf{y}_E) = \int_{\mathbf{x}} F_{S,E}(\mathbf{y}_E|\mathbf{x}) f_x(\mathbf{x}) d\mathbf{x} \qquad (3.25)$$

where again the index E refers to extreme values. It is anticipated that the distribution function on the left-hand side can also be approximated by a parameterized Gumbel-type of probability model.

Similar to the scalar case, the most consistent approach for establishing the long-term extreme value distribution is by integrating the outcrossing-rate across the basic environmental parameters:

$$F_{L,E}(\mathbf{r}_E) = \exp\left\{ -\left(T \int_{\mathbf{x}} v^+(\mathbf{r}_E|\mathbf{x}) f_x(\mathbf{x}) d\mathbf{x} \right) \right\} \qquad (3.26)$$

However, it is important to note that for the multivariate case this cannot be performed *unless the shape of the safe domain has been specified in advance*.

Returning again to the simplified "monochromatic analysis" it is relevant to make a comparison between different types of bivariate Gumbel distributions. Here, a summary of the more detailed analysis performed in Leira and Myrhaug [39], is given. The study is concerned with a comparison of a new transformation-type model and a classical Gumbel Type A bivariate model.

The joint distribution function for the Gumbel Type A distribution as formulated by Gumbel is expressed in terms of the respective marginal distributions on the following form:

$$F_{Z_1 Z_2}(z_1, z_2) = F_{Z_1}(z_1) F_{Z_2}(z_2) \exp\left[-\frac{\theta}{\left\{ \left(\frac{1}{\ln\{F_{Z_1}(z_1)\}} \right) + \left(\frac{1}{\ln\{F_{Z_2}(z_2)\}} \right) \right\}} \right] \qquad (3.27)$$

The role of the parameter θ in this expression is to assign correlation between the basic variables.

Gumbel also described a second type of bivariate extreme value distributions (Type B), and several other types of bivariate extreme-value distributions are found in the literature, see e.g., Johnson and Kotz [13]. However, these models are not considered in any detail in the following.

A comparison between the joint probability density function for the Gumbel Type A and a transformation-based model is shown in Fig. 3.5. Normalized variables are applied in order to make the comparison as general as possible.

The comparison is performed for a correlation coefficient of 0.6. The respective correlation parameters for the two Gumbel distributions are selected such that the correlation coefficient will be the same. The correlation coefficient is evaluated by

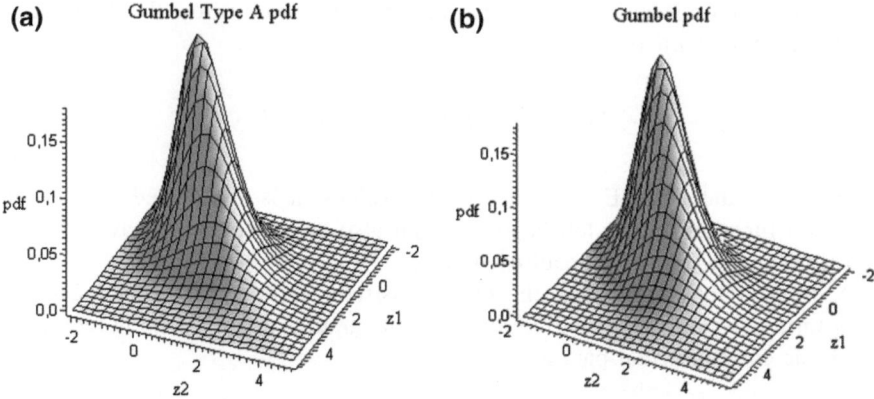

Fig. 3.5 Comparison between joint pdf for Gumbel Type A ($\theta = 0.94$) and transformation-based model. Correlation coefficient $\rho = 0.6$. **a** Gumbel Type A pdf, **b** Bivariate transformation-based Gumbel pdf

numerical integration for both types of models. It is found that for the Type A model the corresponding value of the parameter θ is 0.94.

In Fig. 3.6, the corresponding iso-level contours are compared. It is observed that the iso-contours for the Type A model (with $p_{max} = 0.18$) are somewhat "less rounded" than for the transformation-based model (also with p_{max} around 0.18).

The covariation field of a bivariate distribution is defined as follows (see Leira [40, 41] and Leira and Myrhaug [39]):

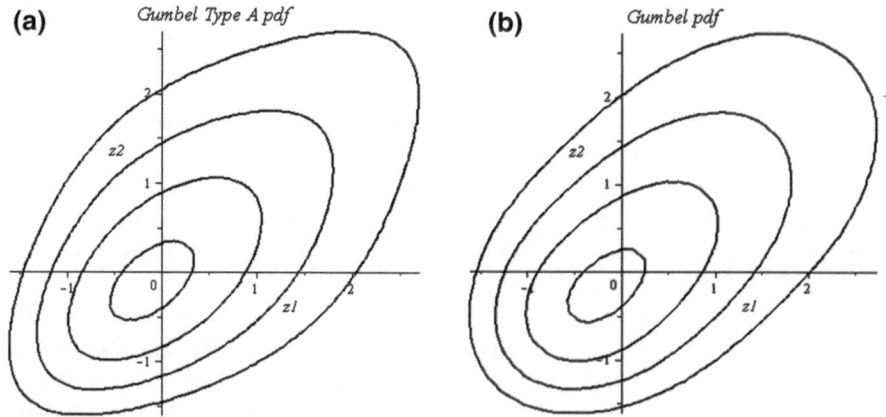

Fig. 3.6 Comparison between iso-contours for the Gumbel Type A pdf and those of the transformation-based model. Iso-contour levels are [0.02, 0.05, 0.1, 0.16] from outer to inner contours for both cases. **a** Gumbel pdf Type A. **b** Transformation-based bivariate Gumbel model

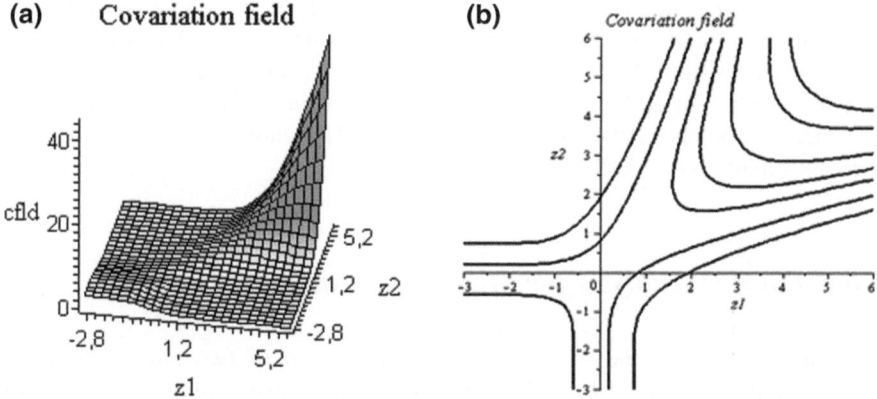

Fig. 3.7 Covariation field for transformation-based model **a** Surface plot. **b** Iso-contour levels (inwards) [0.5, 1., 2., 3., 5., 10., 15.]

$$\text{Covarf}(z_1, z_2) = \frac{p(z_1, z_2)}{p(z_1) \cdot p(z_2)} \tag{3.28}$$

where $p(z_1, z_2)$ is the joint density function for the correlated case and $p(z_1)$ and $p(z_2)$ are the two marginal density functions. This field carries information about the "local covariation structure" of the basic variables at each point.

A comparison between the covariation fields which correspond to the two different models is provided by Figs. 3.7 and 3.8. In parts 3.6a and 3.7a the surface

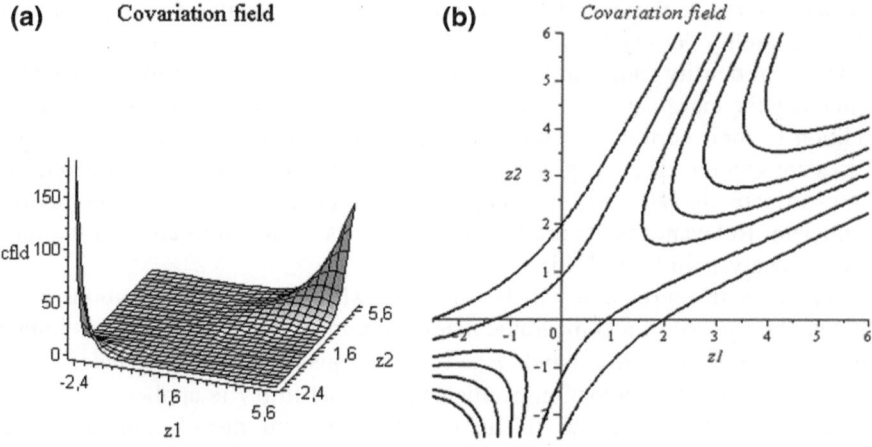

Fig. 3.8 Covariation field for Gumbel Type A model **a** Surface plot. **b** Iso-contour levels (inwards) [0.5,1.,2.,3.,5.,10.,15.]

plots are shown, while parts 3.6b and 3.7b show the corresponding iso-contour plots. It is seen that the maximum value for the Gumbel type A model is significantly higher within the quadrant for which both Z_1 and Z_2 are negative.

Furthermore, for both types of Gumbel models a ridge behavior is observed for intermediate values of the two basic variables. The values along the ridge also increase rapidly towards the outer extensions.

3.6.4 Fatigue Analysis for Multi-Dimensional Processes

The fatigue damage due to a nonlinear combination of two stress component processes, each with a Weibull long-term cycle distributions is subsequently considered. The resulting fatigue damage which is accumulated during a time period T for a one-slope SN-curve (which is expressed on logarithmic form as $\log N = \log \bar{a} - m \log \Delta \sigma$) is obtained as:

$$
\begin{aligned}
E[D(T)] &= \frac{N(T)}{\bar{a}} E[(\Delta \sigma)^m] \\
&= \frac{N(T)}{\bar{a}} \int_0^\infty \int_0^\infty \left[\sigma_{eq}(x_1, x_2)\right]^m f_{X_1 X_2}(x_1, x_2) dx_1 dx_2 \\
&= \frac{N(T)}{\bar{a}} \int_0^\infty \int_0^\infty \left[\sqrt{x_1^2 + c x_2^2}\right]^m f_{X_1 X_2}(x_1, x_2) dx_1 dx_2
\end{aligned}
\tag{3.29}
$$

where N(T) is the number of stress cycles that occur during the period T, which is here assumed to be the same for both components. The quantity $f_{X_1 X_2}(x_1, x_2)$ is the joint probability density function of the two stress cycle processes.

The joint density function is given as a bivariate Weibull distribution which is defined by the marginal shape and scale parameters for each of two components in addition to their correlation structure. When the "intensities" of the stress processes are specified (i.e., as given in terms of their scale parameters), it is relevant to investigate the effect of varying the shape parameters and the correlation coefficient. Relevant results are summarized in the following based on the more detailed analysis in Leira [42].

The effect of correlation will be strongest when the two contributions to the fatigue damage are of a comparable magnitude. In a statistical sense, this implies that the two quantities $E(x_1^2)$ and $cE(x_2^2)$ are set equal to each other. Hence, a base case value of the "combination coefficient" c equal to 0.7 is applied.

Results for two different SN-curves with exponent m = 5 and m = 3 are compared. Two cases with shape factor 0.5 and 2.5 are studied. The shape factor is taken to be the same for both stress components. The scale parameters for the two components are $s_1 = 1.1$, $s_2 = 1.32$. Two different correlation coefficients between them are considered, i.e., $\rho = 0.5$ and $\rho = 0.8$.

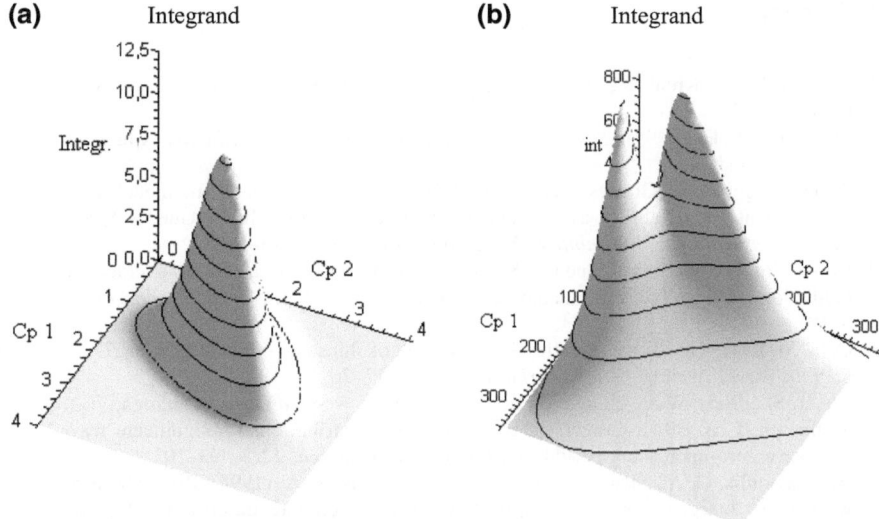

Fig. 3.9 Integrand for example case with $\rho = 0.5$. Fatigue SN-exponent = 5. **a** Shape parameter is 2.5. **b** Shape parameter is 0.5

Table 3.1 Comparison of results for different cases

Shape parameter	0.5		2.5	
Correlation	0.	1.0	0.	1.0
SN-exponent m = 3	63.6	120.0	0.17	0.18
SN-exponent m = 5	3.84×10^5	1.46×10^6	0.58	0.81

Fatigue damage is normalized by a value of 0.0095

An example of the "geometry" of the integrand which is to be evaluated (i.e., for the double integral in Eq. (3.29)) is shown in Fig. 3.9 for a shape parameter of 2.5 (left) and 0.5 (right). The correlation coefficient is 0.5 and the fatigue exponent is $m = 5$. As observed, the shapes are very different for the two different shape parameters.

The results which are obtained for different combinations of the shape parameter, the correlation coefficient and the SN-curve exponent are summarized in Table 3.1.

It is observed that the estimated fatigue damage varies strongly as a function of the fatigue exponent. This applies in particular for the case with a shape parameter of 0.5. Furthermore, the maximum effect of correlation occurs for the combination with $m = 5$ and shape parameter 0.5. For this case the fatigue damage triples when going from zero to full correlation.

References

1. Madsen, H. O., Krenk S., & Lind N. C. (1986). *Methods of structural safety*. New Jersey: Prentice-Hall.
2. Melchers, R. E. (1999). *Structural reliability—analysis and prediction*. Chichester, UK: Ellis-Horwood Ltd.
3. Bitner-Gregersen E., Guedes Soares C. (1997). Overview of probabilistic models of the wave environment for reliability assessment of offshore structures. In C. Guedes Soares (Ed.), *Advances in safety and reliability*, Pergamon (Vol. 2, pp. 1445–1456).
4. Bitner-Gregersen, E., & Guedes Soares, C. (2007). Uncertainty of average steepness prediction from global wave databases. *Proceedings of MARSTRUCT, Glasgow, UK* (pp. 3–10).
5. Ochi, M. K. (1978). Wave statistics for the design of ships and ocean structures. *Transactions Socialis Naval Architects and Marine Engrs, 60*, 47–76.
6. Haver, S. (1985). Wave climate off northern Norway. *Applied Ocean Research, 7*(2), 85–92.
7. Mathiesen, J., & Bitner-Gregersen, E. (1990). Joint distributions for significant wave height and wave zero-upcrossing period. *Applied Ocean Research, 12*(2), 93–103.
8. Athanassoulis, G. A., Skarsoulis, E. K., & Belibassakis, K. A. (1994). Bivariate distributions with given marginals with an application to wave climate description. *Applied Ocean Research, 16*, 1–17.
9. Ferreira, J. A., & Guedes Soares, C. (2001). Modelling bivariate distributions of significant wave height and mean wave period. *Applied Ocean Research, 24*, 31–45.
10. Prince-Wright, R. (1995). Maximum likelihood models of joint environmental data for TLP design. In: C. Guedes Soares et al (Eds.), *Proceedings of the 14th International Conference on Offshore Mechanics and Arctic Engineering, ASME, New York* (Vol. 2, pp. 535–445).
11. Soares, C. S., & Guedes Soares, C. (2007). Comparison of bivariate models of the distribution of significant wave height and peak wave period. *Proceedings of OMAE 2007, San Diego, USA*.
12. Nerzic, R., & Prevosto, M. (2000). Modelling of wind and wave joint occurence probability and persistence duration from satellite observation data. *Proceedings of the 10th International Offshore and Polar Engineering Conference, Seattle, USA*, (Vol. 3, pp. 154–158).
13. Johnson, N. L., & Kotz, S. (1972). *Distributions in statistics: continuous multivariate distributions*. New York: Wiley.
14. Fouques, S., Myrhaug, D., & Nielsen, F. G. (2004). Seasonal modelling of multivariate distributions of met ocean parameters with application to marine operations. *Journal of Offshore Mechanics and Arctic Engineering, 126*, 202–212.
15. Jonathan, P., Flynn, J., Ewans, K. (2010). Joint modelling of wave spectral parameters for extreme sea states. *Ocean Engineering*, doi:10.1016/j.oceaneng.2010.04.004.
16. Clough, R., & Penzien, J. (1975). Dynamics of structures. McGraw-Hill.
17. Newland, D. E. (1993). An introduction to random vibrations (3rd ed.). Longman Scientific & Technical.
18. Fossen, T. I. (2002). Marine control systems. Guidance, navigation and control of ships, rigs and underwater vehicles, Marine Cybernetics, Trondheim, Norway.
19. Newmark, N. M. (1959). A method of computation for structural dynamics. Journal of Engineering Mechanical Division ASCE, 85, EM3, 67–94.
20. Belytschko, T., & Schoeberle, D. F. (1975). On the unconditional stability of an implicit algorithm for nonlinear structural dynamics. *Journal of Applied Mechanics, 97*, 865–869.
21. Hughes, T. J. R. (1976). Stability, convergence and decay of energy of the average acceleration method in nonlinear structural dynamics. *Computers and Structures, 6*, 313–324.
22. Hughes, T. J. R. (1977). Note on the stability of new marks algorithm in nonlinear structural dynamics. *International Journal of Numerical Methods in Engineering, 11*, 383–386.
23. Bathe, K.-J. (1996). *Finite element procedures*. Englewood Cliffs, NJ: Prentice Hall PTR.

24. Krenk, S. (2008). Extended state-space time integration with high-frequency energy dissipation. International Journal for Numerical Methods in Engineering, *73*, 1767–1787.
25. Risken, H. (1989). *The Fokker-Planck equation* (2nd ed.). Berlin: Springer.
26. Kumar, P., & Narayanan, S. (2006). Solution of Fokker-Planck equation by finite element and finite difference methods for nonlinear systems, Sādhanā, (Vol. 31, Part 4, pp. 445–461).
27. Masud, A., & Bergman, L. A. (2005). Solution of the four dimensional Fokker-Planck equation: still a challenge. *Proceedings of ICOSSAR 2005, Millpress, Rotterdam*, ISBN 90 5966 040 4.
28. Rice, S. O. (1944). Mathematical analysis of random noise. *Bell System Technical Journal*, *23*, 282–332 and *24*, 46–156.
29. Longuet-Higgins, M. S. (1952). On the statistical distribution of the heights of sea waves. Journal of Maritime Research, *11*(3).
30. Cartwright, D. E., & Longuet-Higgins, M. S. (1956). On the statistical distribution of the maxima of a random function. *Proceedings of the Royal Society of London*, (Vol. A237, pp. 1706–1711).
31. Wen, Y. K., & Chen, H. C. (1989). System reliability under time varying loads, Part I and II. *Journal of Engineering Mechanics ASCE, 115*(4), 808–839.
32. Hagen, Ø., & Tvedt, L. (1991). Vector process out-crossings as parallell system sensitivity measure. *Journal of Engineering Mechanics ASCE, 117*(10), 2201–2220.
33. Beck, A. T., & Melchers, R. E. (2004). On the ensemble crossing rate approach to time variant reliability analysis of uncertain structures. *Probabilistic Engineering Mechanics, 19*, 9–19.
34. Almar-Naess, A. (1999). Fatigue handbook: Offshore steel structures 3rd revision, Tapir Forlag, Trondheim.
35. Schall, G., Faber, M. H., & Rackwitz, R. (2001). Ergodicity assumption for sea states in the reliability estimation of offshore structures. *Journal of Offshore Mechanics and Arctic Engineering 123*(3), 241, 246.
36. Leira, B. J., Larsen, C. M., Meling, T. S., Berntsen, V., Stahl, B., & Trim, A. (2005). Assessment of fatigue safety factors for deep-water risers in relation to VIV. *Journal of OMAE, 127*(353), 358.
37. Veneziano, D., Grigoriu, M., & Cornell, C. A., (1977). Vector-process models for system reliability. *Journal of Engineering Mechanics Division ASCE, 103*(EM3), 441–460.
38. Gupta, S., & Manohar, C. S. (2005). Multivariate extreme value distributions for random vibration applications. *Journal of Engineering Mechanics ASCE.*
39. Leira, B. J., & Myrhaug, D. (2011). Multivariate gumbel distributions for reliability assessment. *Procdings of OMAE 2011 Conference, Rotterdam.*
40. Leira, B. J. (2010a). A comparison of some multivariate weibull distributions. *Proceedings of 29th OMAE Conference, Shanghai, China.*
41. Leira, B. J. (2010b). Some multivariate weibull distributions with application to structural reliability assessment. *Proceedings ESREL Conference, Rhodes, Greece.*
42. Leira, B. J. (2011). Probabilistic assessment of weld fatigue damage for a non-linear combination of correlated stress components. *Probabilistic Engineering Mechanics, 26*(3), 492–500.

Chapter 4
Categories of On-Line Control Schemes Based on Structural Reliability Criteria

Abstract The present chapter summarizes some of the main principles that can be applied in order to establish on-control schemes based on structural reliability measures. These principles are illustrated in connection with a scalar response process that is composed of a low-frequency part and a superposed high-frequency part. The low-frequency response component is modified by the control algorithm while the high-frequency component is left unchanged. Subsequently, it is discussed in more detail how the control coefficient which is to be applied as part of a LQG control scheme can be calibrated by means of structural reliability methods. Further and more realistic examples of application of the different methods are given in the next Chapter.

Keywords Reliability measures · On-line control schemes · Calibration of LQG

4.1 General

Control schemes are typically based on minimization of objective functions (or loss functions). These are of different types depending on the specific control algorithm to be applied. Frequently, the loss functions are expressed in terms of costs associated with the response processes and/or the control processes. It is rarely the case that structural reliability criteria and the associated cost of structural failure are explicitly taken into account.

In the present chapter, a summary of some approaches for incorporation of structural reliability criteria into the control algorithm is first given. Subsequently, the particular case of Linear Quadratic Gaussian (LQG) control is considered for a simplified system with quasi-static response behavior (which implies that dynamic amplification effects can be disregarded). A comparison is made between the losses which are obtained by application of the traditional objective function versus an objective function that includes the costs associated with structural failure. A second case where the slowly varying component of the response is controlled by LQG and the rapidly varying component is left unchanged is also

B. J. Leira, *Optimal Stochastic Control Schemes Within*
a Structural Reliability Framework, SpringerBriefs in Statistics,
DOI: 10.1007/978-3-319-01405-0_4, © The Author(s) 2013

considered. This will be relevant e.g. in connection with dynamic position control of marine structures.

In relation to formulation of structural response criteria within a specific control scheme, there are at least three possible approaches. The first one corresponds to off-line determination of optimal values of specific control parameters based on structural criteria. These values are subsequently applied as part of the implemented control scheme. The second approach is based on monitoring the values of specific structural reliability measures which are estimated from the observed response processes. If these measures exceed certain pre-defined values, a control action is activated. Alternatively, a switch can be made to a different control scheme if an initial control algorithm has already been activated. The third approach corresponds to making the control action at each time step being continuously updated as a function of one or more structural reliability measures. These three different schemes are discussed in the following.

4.2 Control Schemes Involving Structural Reliability Criteria

4.2.1 Introduction

In the following, focus is on dynamic structural response and control schemes which apply to a sequence of stationary conditions. The external excitation is assumed to contain a high-frequency component which is superposed on a slowly varying load component. This will e.g. be the case for structures which are subjected to low-frequency wind or wave forces in addition to first-order wave loading. In the following, control schemes which make explicit use of structural reliability criteria are considered.

A review of some relevant control schemes based on structural reliability criteria is first given. These schemes are classified as (1) off-line schemes (2) reliability monitoring schemes and (3) on-line schemes as discussed above. The first category is considered in more detail in the present Chapter. The two other categories are discussed here in connection with a simplified example, while more realistic and detailed examples of application are given in Chap. 5.

4.2.2 Various Types of Simplified Structural Reliability Indices

Some relevant categories of simplified reliability measures are considered in the following with focus on scalar response processes. The failure probability for this case was expressed in Chap. 2 by an integral which involves the cumulative distribution function of the capacity variable, $F_R(r)$, and the density function of the load-effect variable, $f_S(s)$. The same probability can alternatively be expressed in terms of the cumulative distribution of the load-effect, $F_S(s)$, and the density function of the resistance, $f_R(r)$, i.e.:

$$p_f = P(Z = R - S \leq 0) = \int_{-\infty}^{+\infty} F_R(x)f_s(x)dx = \int_{-\infty}^{+\infty} [1 - F_S(x)]f_R(x)dx \quad (4.1)$$

For stochastic response processes, the cumulative probability distribution function in Eq. (4.1) will generally refer to the extreme value distribution function for a specific duration. For a Gaussian response process, this distribution function was defined in Chap. 3, Eqs. (3.9–3.14). The probability of failure for this case is accordingly expressed as:

$$
\begin{aligned}
p_f &= \int_{-\infty}^{+\infty} \left[1 - F_{X_{E,T}}(x)\right]f_R(x)dx \\
&= \int_{-\infty}^{+\infty} [1 - \{\exp[-(\exp\{-\alpha(x - u)\})]\}]f_R(x)dx
\end{aligned}
\quad (4.2)
$$

where the parameters α and u are defined in Chap. 3.

If additional sources of uncertainty are taken into account, a multidimensional integration will result as discussed in Chap. 2. Although simplified methods for evaluation of the failure probability exists, already the integral in Eq. (4.2) will require some computational efforts if it needs to be evaluated a large number of times. Hence, this motivates for consideration of simplified reliability measures.

For the particular case that the statistical scatter associated with the resistance can be neglected, the integration is avoided and the following expression is obtained:

$$p_f = \left[1 - F_{X_{E,T}}(r_{th})\right] = [1 - \{\exp[-(\exp(-\alpha(r_{th} - u)))]\}] \quad (4.3)$$

where r_{th} is the given threshold value of the resistance at which failure occurs (e.g. yield stress or breaking stress).

Having computed the failure probability, the so-called reliability index can subsequently be obtained by inverting the standard cumulative distribution function, $\Phi(\cdot)$, for that particular probability level, i.e.:

$$\beta = -\Phi^{-1}(p_f) \quad (4.4)$$

For the case with a fixed value of the threshold capacity the expression in Eq. (4.3) is applied, i.e.

$$\beta = -\Phi^{-1}(p_f) = -\Phi^{-1}\left(\left[1 - F_{X_{E,T}}(r_{th})\right]\right) \quad (4.5)$$

For the case that the response process is Gaussian, the parameters which are required in Eqs. (4.2) and (4.3) are readily obtained. However, if the response process is non-Gaussian (e.g. due to a low-frequency non-Gaussian component) the estimation of the extreme-value distribution becomes more cumbersome. For the case that both a low-frequency and a wave frequency response component is present, estimation of the combined extreme value is not generally a trivial task.

Due to the computational efforts which are required in general, it may not come as a surprise that simplified expressions for the reliability index is in demand. Furthermore, for derivation of on-line control schemes such simplifications make the analysis far easier.

Some of the possible options are summarized in the following, and the strong versus weak features are discussed in each case. For all these indices, the failure probability can be directly estimated by means of the cumulative distribution function if the Gaussian distribution is assumed to apply (which is frequently a very dubious assumption as seen from the discussion above).

From Chap. 2 it is seen that the reliability index for the case that both the load effect and the resistance are Gaussian, the reliability index is expressed as follows (see Eq. (2.10):

$$\beta = \frac{\{\mu_R - \mu_S\}}{\sqrt{\sigma_R^2 + \sigma_S^2}} \tag{4.6}$$

where μ_R, μ_S are the mean values and σ_R^2, σ_S^2 are the variances of the two Gaussian variables R and S.

The first version of a very simplified expression for the reliability index can then be obtained by neglecting the first term in the square-root sign which gives a "time-invariant structural reliability index". This index is obtained as:

$$\beta_{\text{simp}} = (r_{\text{th}} - E[s])/\sigma_s \tag{4.7}$$

where the mean value of the load effect, E[s], will be varying with time if a low-frequency response component is present. Clearly, the present index becomes inaccurate if the variability of the strength (i.e. σ_R^2) is significant as compared to the variability of the high-frequency component of the load effect, i.e. σ_S^2. Another weakness is that this index does not depend on the duration of the period which is considered. Accordingly, the extreme response level for that duration is not properly accounted for.

A more rapidly varying index is obtained by applying the instantaneous value of the response process, s(t), rather than the slowly varying mean value. It is also relevant to apply the standard deviation of the strength variable instead of the load effect (as the latter is more difficult to estimate). This "instantaneous structural reliability index" is accordingly expressed by,

$$\beta_{\text{inst}} = (r_{\text{th}} - s(t))/\sigma_R \tag{4.8}$$

For this case the extreme response level extrapolated into the near future is not properly reflected, i.e. the extreme response is only captured at the time instant that it occurs. This implies that the control action at that time may not be adequate e.g. due to transient dynamic effects.

Proper representation of the extreme value is part of the motivation for introduction of the "delta-index". The extreme response level for a certain extrapolated reference duration is then estimated by multiplying the standard deviation by a "gust" factor, k. For a Gaussian process, this factor is expressed as a function of the duration of a specified time interval [through the number of local maxima that are expected to occur during that period, see Eq. (3.14)]. The delta index is now defined as:

$$\delta = \frac{\left\{ r_{th} - \left(E[s] + k\sigma_{S,HF} \right) \right\}}{\sigma_R} \tag{4.9}$$

where $\sigma_{S,HF}$ is the standard deviation of the high-frequency response component. Also for this case the standard deviation of the strength variable, i.e. σ_R, has been applied in the denominator for simplification purposes.

If required, an extended delta index can also be defined by replacing σ_R with the more complete expression, i.e.:

$$\delta_{Mod} = \frac{\left\{ r_{th} - \left(E[s] + k\sigma_{S,HF} \right) \right\}}{\sqrt{\sigma_R^2 + \sigma_{S,HF}^2}} \tag{4.10}$$

This index is more "accurate" but may also be more challenging in relation to derivation of proper control schemes and assessment of corresponding stability properties.

The positive and negative features associated with the various types of indices are summarized in Table 4.1.

In the following section a comparison is made between the "behavior" of the different indices in relation to a simplistic example. Further examples of application in connection with specific types of control schemes are given in Chap. 5, where different indices are applied as particular features are in focus for each case.

Still other types of indices are also possible. Based on the observations in the present table, it appears that the delta-index and the extended delta-index are to be preferred in relation to reflecting the extreme-value properties of the response process.

Clearly, there is room for creating a uniform approach that can be adapted to each application based on certain criteria. However, we note that the different indices are quite related as some of them can be obtained by a "conversion" is the others. As an example, the conversion between the delta-index and the extended delta-index can be performed by just scaling the denominator in a proper way.

Table 4.1 Comparative features of different types of structural reliability indices

Reliability index type	Positive features	Negative features
$\beta = -\Phi^{-1}(p_f)$ Eqs. (4.4) and (4.5)	Most "correct" index if p_f is computed in a proper way	Computationally demanding
		Difficult to apply for derivation of on-line control schemes
		Neglects variability associated with strength parameter if the version in Eq. (4.5) is applied
Time-invariant index	Very simple to calculate	Does not reflect extreme response levels properly
	Relatively simple to apply for derivation of control schemes	Neglects variability associated with strength parameter
Instantaneous index	Very simple to calculate	Does not reflect extreme response levels properly
	Relatively simple to apply for derivation of control schemes	Neglects variability associated with load effect
	Accounts for variability of strength parameter	
Delta-index	Reflects extreme response levels "within the near future"	May require on-line estimation of extreme-value distribution
	Accounts for variability of strength parameter	Does not reflect variability of load effect in a complete way
	Relatively simple to apply for derivation of control schemes	
Extended delta-index	Reflects extreme response levels "within the near future"	May require on-line estimation of extreme-value distribution
	Accounts for variability of strength parameter	Difficult to apply for derivation of on-line control schemes

4.2.3 Simplistic Illustrative Example and Comparison of Reliability Indices

In order to illustrate the differences between the various types of reliability indices a simplistic example is applied. A stochastic response process is considered which is composed of a (deterministic) low-frequency component and a Gaussian wave frequency component. The former component has a constant (dimensionless) amplitude of 3 and has a period of 80 s. The Gaussian component is defined by a triangular spectral density with a peak of magnitude 1.0 at a period of 6 s. The symmetric spectral density becomes zero at 4 and 8 s.

Figure 4.1 shows the wave frequency response component, the low-frequency component and the sum of the two components (in the upper, middle and lower part of the figure). The expected number of local maxima for the high-frequency component corresponding to a duration of 600 s (i.e. 10 min) is $N = 100$. The standard deviation of the wave frequency component is $\sigma_{r,HF} = 0.31$, which implies that the expected largest value during 600 s is 1.0.

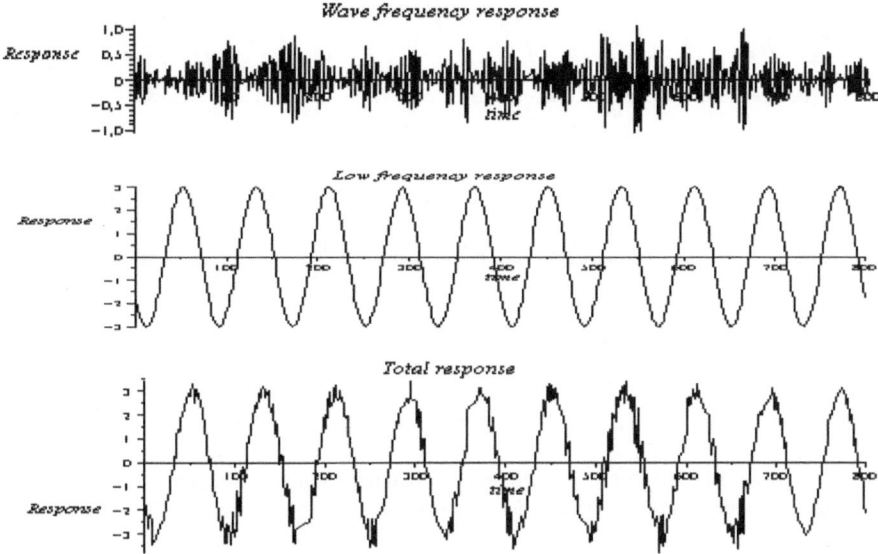

Fig. 4.1 Wave-frequency (*upper*), low-frequency (*middle*) and total response (*lower*)

The total duration of the response process considered below is 800 s which corresponds to 10 periods of the low-frequency component.

For the purpose of illustrating the differences between the various indices, the critical level is set equal to 3.5. The minimum value of the simplified index is accordingly obtained as $(3.5 - 3)/0.31 = 1.6$ and the maximum value becomes $(3.5 + 3)/0.31 = 21$. The variation of this index between these two limits is shown in the upper part of Fig. 4.2.

The instantaneous index captures the variation of the high-frequency response component as seen in the middle part of Fig. 4.2. This index is normalized by the standard deviation of the capacity, which is set to $\sigma_R = 0.5$. The minimum value is now seen to be negative due to subtraction of the high-frequency response component. The maximum positive value is significantly smaller than for the simplified index. This is due to the subtraction of the high-frequency component and normalization by $\sigma_R = 0.5$ instead of by $\sigma_{r,HF} = 0.31$ as for the simplified index.

The variation of the third option, the delta index, is shown in the lower part of Fig. 4.2. The quantity $k \cdot \sigma_{r,HF}$ is set equal to the expected largest value of the high-frequency component during 10 min, i.e. 1.0. It is seen that both the minimum and the maximum values of this index are smaller than for the instantaneous index due to this "amplification" term.

For comparison, the reliability index based on the extreme-value distribution for the high-frequency component is also calculated. For the present case, the low frequency-component needs to be taken into account, and the probability of failure is then expressed as

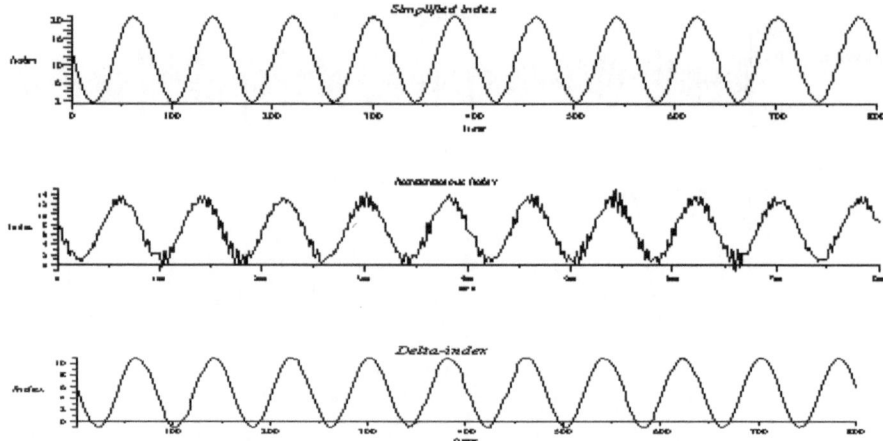

Fig. 4.2 Time variation of three different reliability indices. Simplified index (*upper*), Instantaneous index (*middle*), Delta index (*lower*)

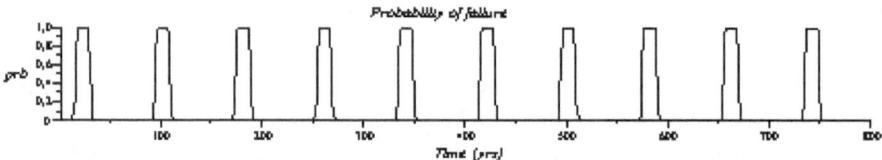

Fig. 4.3 Probability of failure based on cumulative Gumbel distribution function

$$p_f = \left[1 - F_{x_{E,T}}(r_{th})\right] = \left[1 - \{\exp[-(\exp(-\alpha(r_{th} - x_{LF}(t) - u)))]\}\right] \qquad (4.11)$$

where $x_{LF}(t)$ is the time-varying low-frequency response component. The variation of this failure probability which forms the basis for calculation of the index is shown in Fig. 4.3. It is seen that it varies between 0 and 1. The corresponding reliability index will then oscillate between plus and minus infinity (and is accordingly not shown in the figure). Hence, for the present response process there is clearly the need for a modifying control action in order to reduce the resulting maximum value of the failure probability.

4.2.4 Off-Line Control Schemes

The different types of relevant control schemes based on application of these reliability indices are next considered. The first category which was referred to above is based on pre-calibrating the parameters of existing and widely applied control schemes such as LQG and PID algorithms. It may be also relevant to modify the "classical" schemes themselves by introducing alternative optimality criteria. As one of several examples related to this category, Tomasula et al. [1] determined the control factor based on a heavier penalty for the high response levels than the traditional quadratic loss function.

In Sect. 4.3 below, a more systematic assessment is made of the inherent failure probability associated with a pure LQG scheme for a quasistatic type of response behavior. The simplified example already discussed above will also be applied for the purpose of illustrating the procedure. Hence, this category of control schemes is not discussed in any further detail at present.

4.2.5 Control Scheme Based on Reliability Monitoring

By monitoring the response level and computing the reliability measure at each time step (e.g. one of the reliability indices above), no control action is activated until a pre-specified threshold level of the index is being exceeded. As soon as this level is being reached the controller is being activated. Furthermore, it is reasonable that the control action is higher the more the threshold value of the index is being exceeded. A control algorithm of this type is described e.g. in [2, 3], and a more detailed description of this application is given in Chap. 5.

At present, this type of control scheme is illustrated in connection with the simplified example which was described in Sect. 4.2.3. For this example, it was observed that without any control scheme being implemented the probability of failure (i.e. the probability of exceeding the response threshold of 3.5) was indeed very high.

The instantaneous index is presently applied and the critical (target) level of this index is set to 5. When this level is reached, the control action is activated according to the following formula:

$$u(t) = 3k \left\{ 1 - \frac{\beta_{inst}(t)}{\beta_{t,inst}} \right\} \qquad (4.12)$$

where k is the stiffness of the structure and $\beta_{t,inst}$ is the target value of the instantaneous reliability index. If the term in brackets becomes negative, it is truncated at zero (i.e. at the point in time when the instantaneous value of the index is regained at the target value. The resulting modification of the response is given by the same expression by just removing the factor k.

The control action (i.e. divided by k) together with the modified total response which results by application of this "monitoring" scheme are respectively shown in the upper and middle part of Fig. 4.4. As observed, the control is only activated for limited time intervals. The maximum positive value of the total response is seen to be lower than the critical value of 3.5.

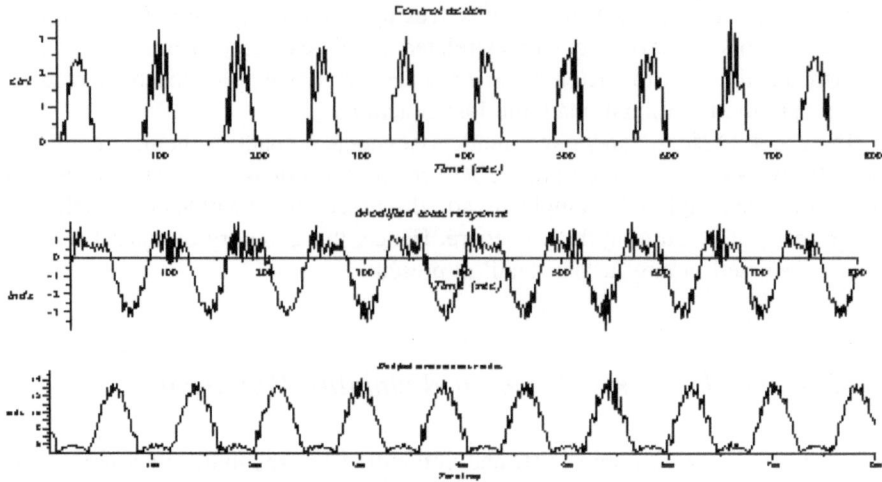

Fig. 4.4 Control action (*upper*), modified total response (*middle*) and modified instantaneous index (*lower*) corresponding to the control scheme based on reliability index "monitoring"

The resulting modified value of the instantaneous index is shown in the lowermost part of Fig. 4.4. It is seen that it stays above the target value of 5 throughout the considered time period.

4.2.6 On-Line Control Schemes

The on-line control scheme is obtained by first introducing a proper objective function (loss function) which is expressed in terms of the relevant reliability index. The same example as above is considered and a quadratic objective function is applied. This function is expressed as follows in terms of the simplified index (assuming that that $E[r]$ is positive):

$$(\beta_{\text{trg}} - ((r_{\text{CR}} - (E[r] - s\Delta u))/\sigma_{r,\text{HF}}))^2 \tag{4.13}$$

where r_{cr} is the (one-sided) critical response level which has a positive value, $E[r]$ is the slowly-varying mean value of the response (i.e. which corresponds to the low-frequency response component), which is modified by means of the incremental control action, Δu; s is a scaling factor which presently is equal to the inverse of the stiffness, i.e. $s = 1/k$. The resulting response modification is hence expressed as $s \cdot \Delta u = \Delta u/k$. The expression for the incremental control action which minimizes the objective function is now found by setting the derivative with respect to this increment equal to zero, which gives:

$$\Delta u = k\{(-\beta_{\text{trg}}\sigma_{r,\text{HF}} + r_{\text{cr}} - E[r])\} \tag{4.14}$$

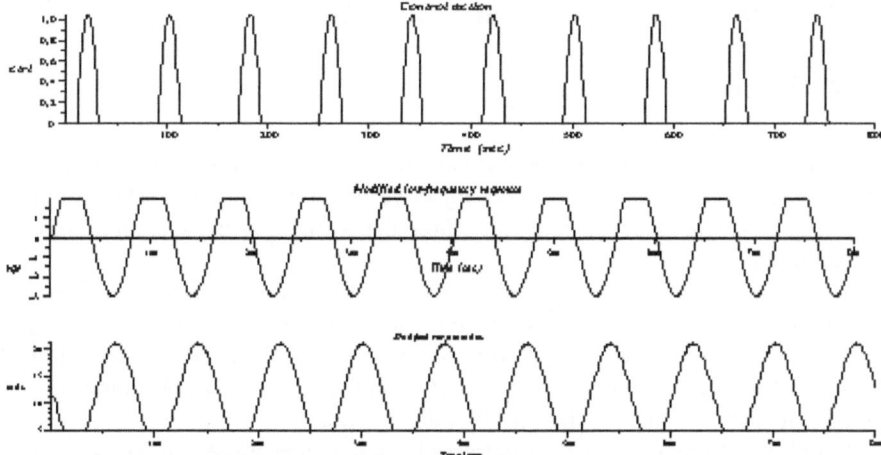

Fig. 4.5 Control action (*upper*), modified low-frequency response (*middle*) and modified value of simple index (*lower*) corresponding to control scheme based on on-line control scheme

for the intervals where E[r] is positive. If E[r] is negative, this term should formally instead be replaced by its absolute value. However, in practical applications the control action should be set equal to zero for this case. The reason is that physically it does not give any meaning to apply energy in order to decrease the reliability index towards the target value if it already has a higher value.

The target value of the reliability index is set to 5 also in this case (however, a somewhat higher value could be contemplated due to the smaller value of the standard deviation for the present index).

The control action (divided by the stiffness k) is shown in the upper part of Fig. 4.5, and the corresponding modified low-frequency response is shown in the middle part of the figure. It is seen that due to the "smooth" behaviour of the present reliability index as compared to the instantaneous index, both the control action and the controlled response become periodic (since the low-frequency response component is periodic). It is seen that the positive part of the low-frequency component is strongly modified.

The resulting modified reliability index is shown in the lower part of Fig. 4.5. As observed, the index stays above the target value of 5 throughout the considered time interval.

Clearly, the objective function for the present case could readily be modified in order to include a term which represents the cost associated with application of the control action. However, the main intention is presently to illustrate how the reliability index can be applied for derivation of an on-line control scheme.

Extension of the present category of control actions to systems with dynamic response effects are illustrated in connection with various more realistic examples in Chap. 5.

4.3 Off-Line "Calibration" of LQG Schemes

4.3.1 General

In the present Section, the particular case of Linear Quadratic Gaussian (LQG) control is considered. The associated objective function is defined. The particular case of quasi-static response behavior where the stiffness of the system dominates is elaborated.

A second "reliability-based" objective function is subsequently introduced which incorporates the costs associated with structural failure. A comparison between the values of the two different objective functions is performed, and it is discussed how the parameters of the LQG control scheme can be tuned to give equivalent results to the "reliability-based" function.

A slightly different case is also addressed where the response consists of the sum of a slowly varying component and a rapidly varying (high-frequency) component. The low-frequency component is controlled by LQG, while the rapidly varying component is left unchanged. This case will be relevant e.g. in connection with position control of marine structures, see e.g. Berntsen et al. [4].

4.3.2 LQG Control of Stationary Quasi-Static Response

The dynamic equilibrium equation also including a control action term can be expressed as

$$m\ddot{x}(t) + c\dot{x}(t) + kx(t) = gu(t) + F(t) \tag{4.15}$$

where the dot superscript designates time differentiation, i.e. $\dot{x}(t)$ is the response velocity and $\ddot{x}(t)$ is the response acceleration; k is the structural stiffness, c is the damping coefficient, m is the mass and g is a scaling factor for the control action u(t). The external loading is denoted by F(t).

For the case of so-called "quasi-static" loading and response, the equilibrium equation is simplified by neglecting the damping and inertia term:

$$kx(t) = gu(t) + F(t) \tag{4.16}$$

The loss function that forms the basis for the LQG control scheme is expressed as:

$$J(x) = E\left[\int_0^T (\alpha x^2 + \beta g^2 u^2)dt\right]\left[\int_0^T (\alpha E[x^2] + \beta g^2 E[u^2])dt\right]$$
$$= T(\alpha E[x^2] + \beta g^2 E[u^2]) \tag{4.17}$$

where α is the proportionality factor related to the response cost and β is the proportionality factor associated with the cost of the control action. The substitution of the factor T for the integration operator is due to the assumption that the load and response processes are stationary processes. Furthermore, the transition from the integral to the non-integral form in Eq. (4.17) implies that transients associated with the initial values are neglected. The weighting of the final state is also the same as for all the intermediate states.

For the case of quasi-static response, this loss function can be further simplified. The control action is then expressed as being proportional to the response level (i.e. neglecting the velocity proportional term which is generally present for LQG-control). This implies that $u = -Cx$ where C is a constant to be determined. For this case, the response can be expressed explicitly in terms of the external excitation by solving the following equation

$$k \cdot x(t) = -gCx(t) + F(t) \tag{4.18}$$

which gives $x(t) = \frac{F(t)}{k+g \cdot C}$. The loss function accordingly becomes

$$J(x) = T\left(\alpha E\left[x^2\right] + \beta g^2 E\left[u^2\right]\right) = T\left(\alpha + \beta g^2 C^2\right) E\left[x^2\right]$$
$$= T\left(\alpha + \beta g^2 C^2\right) \frac{E\left[F^2\right]}{(k+gC)^2} \tag{4.19}$$

where $E[F^2]$ is the second moment of the external excitation force.

Differentiating this function with respect to the control parameter and setting the resulting expression equal to zero, the following is obtained, [5]:

$$\frac{2g(-\beta kgC + \alpha)}{(k+gC)^3} = 0 \tag{4.20}$$

Solving this equation gives the value of the "control coefficient" which minimizes the loss function:

$$C_{Opt} = \frac{\alpha}{\beta kg} \tag{4.21}$$

For simplicity and without loss of generality, the "gain factor" is next taken to be equal to the structural stiffness, i.e. $k = g$. A "control factor" is introduced which scales the optimal value of the control coefficient, i.e. $C = fC_{Opt} = f\frac{\alpha}{\beta k^2}$. By further introducing the cost coefficient ratio $r = \frac{\alpha}{\beta k^2}$ the loss function can be expressed as

$$J(r,f) = TE\left[F^2\right]\left(\frac{\alpha}{k^2}\right)\frac{(1+rf^2)}{(1+rf)^2} = TE\left[F^2\right]\left(\frac{\alpha}{k^2}\right)L(r,f) \tag{4.22}$$

The factor $TE\left[F^2\right]\left(\frac{\alpha}{k^2}\right)$ is subsequently taken to be a fixed value and is removed from the expression. The resulting normalized expression, i.e. L(r,f), is displayed

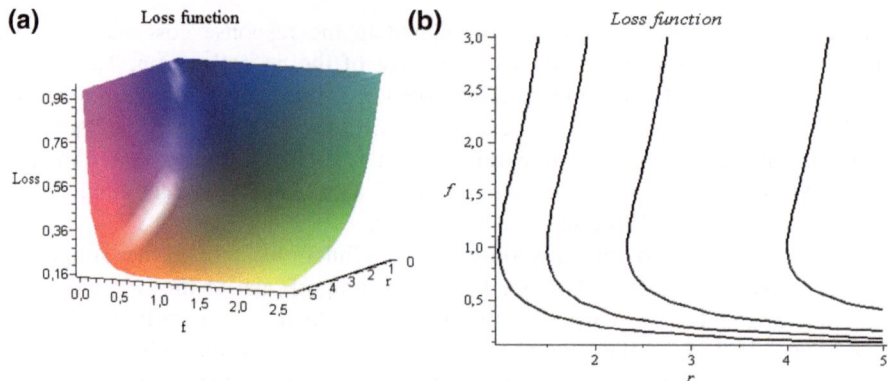

Fig. 4.6 Example of "normalized" LQG loss function for the case that g = k. Contour levels in 4.2 (**b**) from *right* to *left* are 0.2, 0.3, 0.4 and 0.5

in Fig. 4.6a below. The contour lines of the loss function which correspond to the levels 0.2, 0.3, 0.4 and 0.5 (from right to left) are displayed in Fig. 4.6b. It is seen that the minimum value complies with the expression just obtained (i.e. corresponding to f = 1). The loss is monotonously decreasing as a function of the cost coefficient ratio r.

By inserting this optimal value of the control coefficient (i.e. which corresponds to the factor f = 1) into the loss function, the normalized minimum loss is expressed as

$$L(r, 1) = \left(\frac{1}{1 + r}\right) \qquad (4.23)$$

4.3.3 Alternative Loss Function Based on Structural Failure Cost

Focusing on the extreme response levels, it may be considered that there is no cost associated with the structural response until the level is reached where failure occurs. Hence, the expected response cost is expressed as the product of the failure cost, C_f, times the probability that failure occurs for the time period, T, which is considered.

The associated loss function can then be expressed as

$$J(x) = \left(C_f p_f(T) + T\beta E\left[u^2\right]\right) \qquad (4.24)$$

The failure probability for a Gaussian response process (for a given duration, T) can be expressed as the probability that the extreme response level exceeds a (deterministic) critical response threshold, x_{cr}. The asymptotic distribution for the extreme response will be of the Gumbel type.

The failure probability can then be expressed in terms of the complement of the cumulative Gumbel distribution function as

$$
\begin{aligned}
p_f(T) &= 1 - F_Y\left(\frac{x_{cr}}{\sigma_x}\right) \\
&= 1 - \exp\left[-\exp\left\{-\left(\sqrt{2\ln(n)}\right)\cdot\left(\frac{x_{cr}}{\sigma_x} - \sqrt{2\ln(n)}\right)\right\}\right]
\end{aligned}
\tag{4.25}
$$

where n is the expected number of local maxima for a given duration T. This number can be estimated as $n = T/T_z$ where T_z designates the zero-crossing period of the response. In the following a representative value of $n = 1{,}000$ is applied. The ratio $\frac{x_{cr}}{\sigma_x}$ is the (normalized) critical response threshold level, where σ_x is the standard deviation of the response process (also including the effect of the control action). This standard deviation can hence be expressed in terms of the standard deviation of the load process, the system stiffness, the loss coefficient ratio (i.e. the quantity r) and the control coefficient (i.e. f) as:

$$
\frac{x_{cr}}{\sigma_x} = \frac{x_{cr}}{\left[\frac{\sigma_F}{k(1+rf)}\right]} = \frac{x_{cr}k(1+rf)}{\sigma_F}
\tag{4.26}
$$

The critical threshold level can also be expressed as a real-valued factor, R, times the "undisturbed" standard deviation of the response, i.e. for the case that no control action is being applied. This gives:

$$
x_{cr} = \frac{\sigma_F R}{k}
\tag{4.27}
$$

and by inserting this in the expression above we obtain:

$$
\frac{x_{xr}}{\sigma_x} = R(1+rf)
\tag{4.28}
$$

The loss function above can then be rewritten on the following form:

$$
\begin{aligned}
J(x) &= \left(C_f p_f(T,R,f) + T\beta E[u^2]\right) \\
&= \left(C_f p_f(T,R,f) + T\beta C^2 k^2 \frac{E[F^2]}{k^2(1+C)^2}\right) \\
&= \left(C_f p_f(T,R,f) + \alpha f^2 r \frac{TE[F^2]}{k^2(1+fr)^2}\right)
\end{aligned}
\tag{4.29}
$$

where the control action has been expressed in terms of the external force and the control parameter.

The cost associated with structural failure is next expressed as a factor R_f times the second term in the loss function (i.e. $C_f = R_f \dfrac{\alpha TE[F^2]}{k^2}$,) which yields

$$J(x) = \frac{\alpha TE[F^2]}{k^2} \left(R_f p_f(T,R,r,f) + \frac{f^2 r}{(1+fr)^2} \right) \qquad (4.30)$$

This loss function is illustrated in Fig. 4.7 for the case that $R_f = 2$, and $r = 1$.

From the figure it is seen that for a specific value of the critical threshold (i.e. R) the value of the control factor f which minimizes the cost function can be identified in a straightforward way. For low values of this critical threshold, i.e. below 2.0, the required values of f are far above 1.0. For values of the critical threshold which are above 4.0, the required value of f is very low, i.e. far below 1.0.

It is interesting to note that for the case without any control action, the characteristic largest response for a duration which corresponds to n = 1,000 (i.e. number of local maxima) is equal to $\sqrt{2\ln(n)} = \sqrt{2\ln(1000)} = 3.72$. The probability of the normalized response level to exceed this response level for the case without any control action (i.e. for f = 0) decreases rapidly due to the shape of the Gumbel distribution. From the figure it is observed that this is reflected in the cost function also decreasing rapidly above this value. The "ridge behavior" of the loss function hence reflects the rapid variation of the failure probability around this particular response level.

For a certain value of the parameter f, the present loss function will have the same value as the LQG loss function above. This equality will require that the first term of the present loss function is equal to the first term of the LQG loss function:

$$R_f p_f(T,R,r,f) = \frac{1}{(1+fr)^2} \qquad (4.31)$$

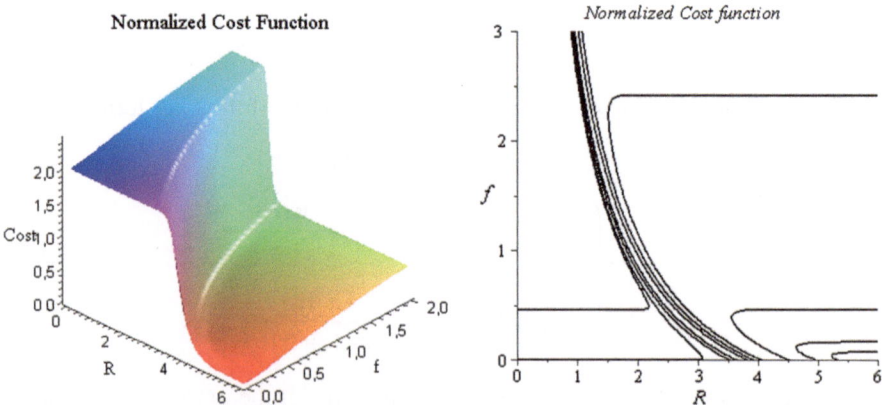

Fig. 4.7 Normalized cost function for $R_f = 2$, r = 1. Contour levels in *lower* part of figure (from *right* to *left*) are (0.005, 0.02, 0.1, 0.5, 0.75, 1.0, 1.5, 1.75, 2.0, 2.1)

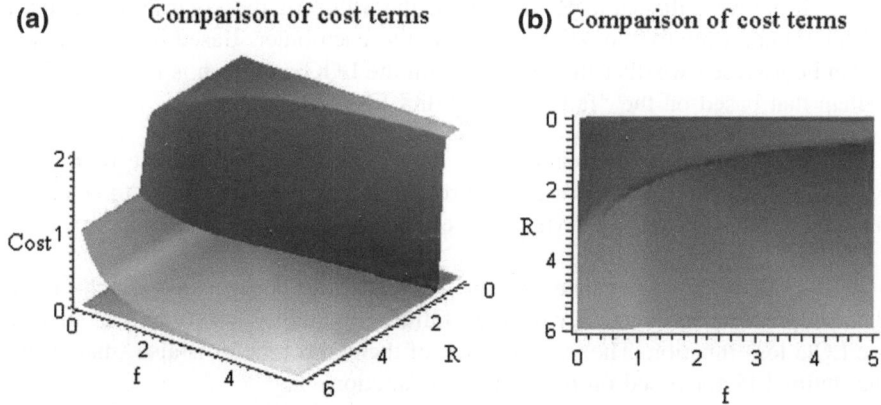

Fig. 4.8 Comparison of cost terms for "failure cost" and LQG quadratic cost terms. (a) *Side* view (b) *Top* view

A comparison of the left and right hand sides of this equation is shown in Fig. 4.8. In part (b) of the figure an aerial view is included in order to show the intersection line, i.e. the collection of simultaneous values of R and f that will make the two loss functions identical to each other.

The intersection curve is located at the "steep wall" of the function on the left-hand side of the equation.

Application of the optimal LQG control parameter corresponds to $f = 1$ which roughly gives the same cost value as for a critical response threshold R of around 2 (for the present value of R_f).

The procedure for calibration of the LQG algorithm by means of the failure probability loss function can e.g. be as follows:

- Identify the β-parameter of the LQG algorithm. Set the initial value of the α parameter equal to $\beta \cdot k^2$ to make the cost coefficient $r = 1$.
- Determine the critical threshold R of the structural system based on mechanical considerations. Determine the failure cost and compute the cost coefficient R_f.
- Determine the optimal value of the dimensionless control parameter f_1 for the "failure probability loss function" based on the critical threshold just obtained.
- Verify that the failure probability which corresponds to f_1 is acceptable.
- Calculate the value of the control coefficient C which corresponds to f_1 (i.e. $C = f_1 C_{Opt} = f_1 \frac{\alpha}{\beta k^2} = f_1 r = f_1$ since $r = 1$).
- Adjust the parameter α for the equivalent LQG scheme such that $\alpha = Ck^2\beta$, i.e. applying the derived value of C from the previous step as the optimal LQG control coefficient. Compute the corresponding updated value of the cost coefficient R_f. If this deviates from the initial value, return to step 2 and iterate until convergence.
- Implement the LQG control scheme based on the resulting value of C (or equivalently the values of α and β).

• For the critical threshold R, determine the value of the control parameter f_2 which makes the two loss functions equal to each other. Based on this value, it can be assessed whether the loss based on the LQG approach is lower or higher than that based on the "failure probability loss function".

   As an example of application, we may consider the case that $R = 3$ and $R_f = 2.0$. The optimal value of the control parameter f_1 is then found to be around 0.67. The failure probability which corresponds to this value of the control factor $r \cdot f = 0.67$ is equal to 0.008.

   The value of f_2 which makes the two loss functions equal is 0.2, which implies that the present loss is smaller for the "failure probability loss function" than for the LQG loss function. The present value of the control factor is also smaller than the optimal factor based on the LQG loss function.

   The reason for starting with a specific value of β is that typically the cost associated with the control action is easiest to estimate. The cost associated with structural response is clearly more dependent on modeling assumptions for the structural system and also associated with some degree of subjective judgment.

4.3.4 The Case with Two-Scale Response Characteristics

Next, we consider the case where the scalar excitation can be expressed as the sum of a low-frequency component, F_L, and a high-frequency component, F_H, i.e. $F = F_L + F_H$. For the case with zero correlation between the two force components, the mean square of the total force becomes equal to the sum of the contributions from the two components. This implies that

$$E[F^2] = E[F_L^2] + E[F_H^2] \qquad (4.32)$$

Similarly, the mean square of the quasistatic response is expressed as the low-frequency and high frequency contributions

$$E[x^2] = E[x_L^2] + E[x_H^2] \qquad (4.33)$$

The control action for this case is only based on the low frequency response component, i.e. $u_L = Cx_L$. The resulting expression for the loss function will then obtain an additional term (designated by h) due to the "uncontrolled" high-frequency component:

$$J(r,f) = TE[F_L^2]\left(\frac{\alpha}{k^2}\right)\frac{(1 + rf^2 + h)}{(1 + rf)^2} \qquad (4.34)$$

where the parameter $h = \dfrac{E[F_H^2]}{E[F_L^2]}$ measures the relative energy contributions from the high-frequency versus the low-frequency components. The relative increase of the loss function is accordingly equal to $h/(1 + rf^2)$. The analysis will then proceed in the same way as for Eqs. (4.19–4.21) just by applying the parameter $\alpha \cdot (1 + h)$ instead of α.

The new optimal value of the control parameter is then obtained as:

$$C_{Opt,LF+HF} = \frac{\alpha(1 + h)}{\beta kg} \tag{4.35}$$

The corresponding modified value of the normalized loss function is accordingly expressed as:

$$L(r, 1) = \left(\frac{1 + h}{1 + r(1 + h)}\right) \tag{4.36}$$

If the high- and low-frequency components are correlated, there will be a second term $h_2 = E[F_H F_L]/E[(F_L)^2]$ which is added to the parameter h in the loss function. For this case, the optimal value of the control parameter will need to be modified due the presence of this correlation term.

For the purpose of illustration, we consider the same simplified example as before for which the low-frequency component is given as a periodic function (i.e. a sine function including a phase angle). The cost function based on the structural failure probability is now modified by incorporation of the low-frequency component in the exponential function, i.e.

$$\begin{aligned}
p_f(T) &= 1 - F_Y\left(\frac{x_{cr}}{\sigma_x}, \frac{x_{LW}}{\sigma_x}\right) \\
&= 1 - \exp\left[-\exp\left\{-\sqrt{2\ln(n)}\right\}\left(\frac{x_{cr}}{\sigma_x} - \frac{x_{LW}}{\sigma_x} - \sqrt{2\ln(n)}\right)\right]
\end{aligned} \tag{4.37}$$

where x_{LW} is the amplitude of the low-frequency response component after being modified by the control action.

The same values corresponding to a low-frequency harmonic component is applied as before with amplitude 3 and period 80 s. Furthermore, the standard deviation of the high-frequency component is still equal to 0.31, and the corresponding term in Eq. (4.36) then becomes:

$$\frac{x_{LW}}{\sigma_x} = \frac{3.0}{0.31}(1 - fr) = 9.68(1 - fr) \tag{4.38}$$

The modified objective function (as compared to the one in Fig. 4.7) based on the cost of structural failure can then be obtained. The result is shown in Fig. 4.9.

The comparison with the cost term for the LQG objective functions will also be different as shown in Fig. 4.10. This figure is to be compared with Fig. 4.8 for the case without any low-frequency component. As observed, the shape and location of the "transition-wall" are now modified.

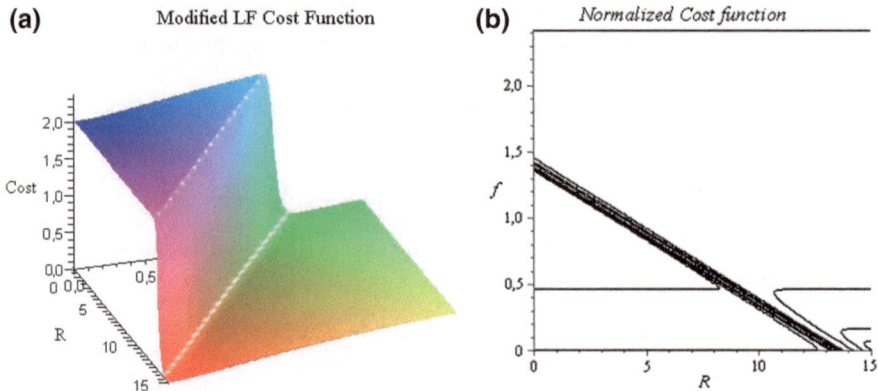

Fig. 4.9 Normalized cost function for n = 1,000. Contour levels in *lower* part of figure (from *right* to *left*) are (0.005, 0.02, 0.1, 0.5, 0.75, 1.0, 1.5, 1.75, 2.0, 2.1)

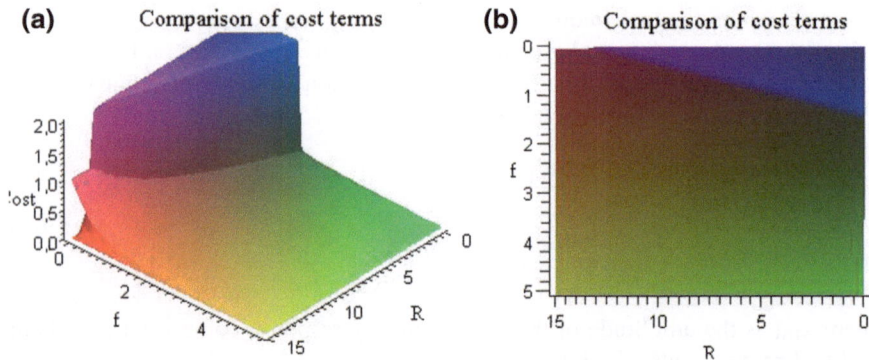

Fig. 4.10 Comparison of cost terms for "failure cost" and LQG quadratic cost terms. Response process is sum of a regular low-frequency component with amplitude 3 and a high-frequency Gaussian component with standard deviation 0.31. (**a**) *Side* view (**b**) *Top* view

4.3.5 Application to Simplified Example

The calibration procedure which was described in the previous sub-section is next applied to the simplified example which was studied in Sect. 4.2. The normalized threshold value for the present example is equal to $\frac{X_{cr}}{\sigma_x} = \frac{3.5}{0.31} = 11.3$. It is found that the optimal control factor, i.e. f, then becomes equal to 0.3. The optimal value of the control coefficient C for the case that r = 1 is then also obtained from Eq. (4.34) as $C_{opt} = 0.3(1 + h)$. For the present example this value is 0.3 (1.02) = 0.306.

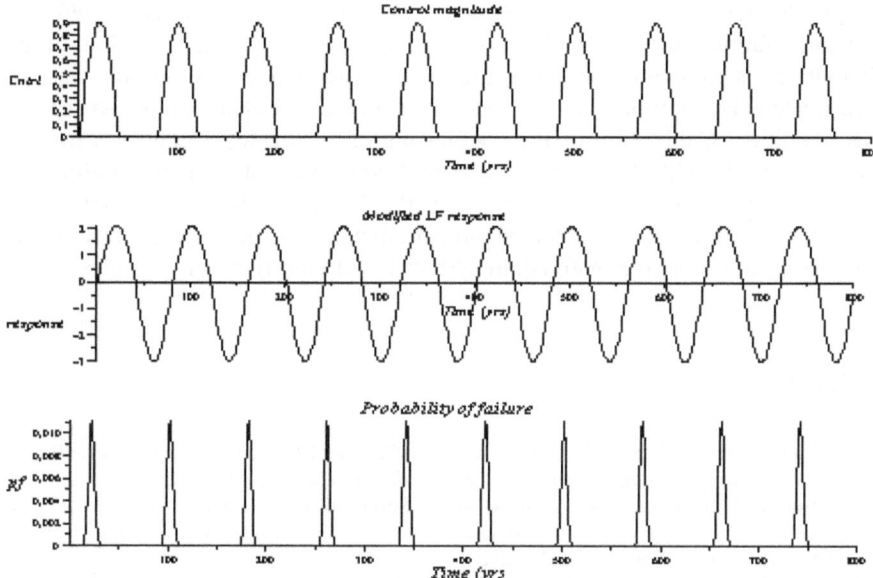

Fig. 4.11 Control action (*upper*), Modified low frequency response (*middle*) and Probability of failure (*lower*) based on optimal control coefficient for "probability-scheme"

The resulting time series for the control action, the modified low-frequency response and the probability of failure are shown in Fig. 4.11.

It is seen that the maximum value of the failure probability is around 0.10 which reflects the ratio between the failure cost and the cost of applying the continuous control action.

4.4 Concluding Remarks

Various approaches for incorporation of structural reliability measured in relation to control procedures were considered. Examples of three different categories of such procedures were discussed.

Subsequently, a specific approach for pre-calibration of the control parameter was considered. A quasistatic structural system was applied in order to clarify the discussion as much as possible. Loss functions associated with Linear Quadratic Gaussian (LQG) control were addressed. These loss functions contain one term which represents the cost associated with the response energy and one term which corresponds to the energy of the control action. The two terms are weighted by their respective cost coefficients. The failure probability which is implied by the relative weighting of these costs was computed (for given values of the critical response threshold).

An alternative loss function which incorporates the cost associated with structural failure due to overload was next introduced. The variation of this alternative loss function with respect to the same relative cost weighting was next investigated. The optimal value of the "control factor" based on this loss function was identified. Calibration of the coefficients of a pure LQG control scheme in order to comply with the associated optimal point was subsequently outlined.

Extension of the comparison between the LQG and failure probability loss functions in relation to cases with significant dynamic response amplification, also comprising multi-degree-of-freedom systems, will clearly be highly relevant.

References

1. Tomasula, D. P., Spencer, B. F., & Sain, M. K. (1996). Nonlinear control strategies for limiting dynamic response extremes. *Journal of Engineering Mechanics, 122*(3), 218–229.
2. Berntsen, P. I. B., Leira, B.J., Aamo, O.M., & Sørensen, A. J. (2004). *Structural reliability criteria for control of large-scale interconnected marine structures.* Proceedings of 23rd OMAE, Vancouver, Canada, 20–25 June.
3. Berntsen, P. I. B., Aamo, O. M.; Leira, B. J., & Sørensen, A. J. (2004). Structural reliability-based control of moored interconnected structures. Control Engineering Practice, 2006, 10.1016/j.conengprac.2006.03.004.
4. Berntsen, P. I. B., Aamo, O. M., & Leira, B. J. (2009). Ensuring mooring line integrity by dynamic positioning: Controller design and experimental tests. *Automatica, 45,* 1285–1290.
5. Leira, B. J. (2011). Structural reliability criteria in relation to on-line control schemes. Proceedings of ICASP, Zurich.

Chapter 5
Example Applications Related to On-line Control Schemes

Abstract In the present chapter examples of application are given for the two last categories of structural reliability-based control schemes. These two categories correspond respectively to on-line monitoring of structural reliability measures and direct implementation of continuous on-line control magnitudes based on the same measures. For both cases on-line evaluation of a structural reliability measure is required. For the former category the examples comprise position control of a floating fish-farm by means of thruster assistance. For the latter category the examples also correspond to position control of floating vessels based on the instantaneous reliability of the attached risers and mooring lines.

Keywords On-line · Control scheme · Structural reliability measures · Monitoring

5.1 General

In the previous chapter, general principles for incorporation of structural reliability criteria as part of control schemes were considered. Three main categories of methods were considered, and the first type which is based on pre-calibration of coefficients to be applied for a specific control scheme was considered in some more detail.

In the present chapter, more detailed examples of application for the two next categories are considered. Both of these categories are based on on-line evaluation of the relevant structural reliability measures. In the first example, the control action is activated when the index exceeds a specified limit. For the other examples, the level of the control action is determined continuously based on the computed value of the structural reliability measure.

Since all the examples are concerned with position control of floating vessels, the dynamic equations describing the motion of such vessels are first reviewed.

B. J. Leira, *Optimal Stochastic Control Schemes Within
a Structural Reliability Framework*, SpringerBriefs in Statistics,
DOI: 10.1007/978-3-319-01405-0_5, © The Author(s) 2013

5.2 Vessel Dynamics

5.2.1 Vessel Motion

Description of the vessel motion is based on the dynamic equilibrium equation including the effect of the control actions. Like conventional marine vessels, a dynamic positioning (DP) vessel is subjected to time-varying environmental loads (waves, wind and current). For a DP vessel, it is not feasible to counteract the wave-frequency (WF) oscillatory response which is caused by first-order loading. The reason is that compensation of the WF components of the motion requires excessive thruster modulation. The control action of the propulsion system of a DP vessel is accordingly activated by the LF part of the vessel movement which is caused by current, wind and second order wave loads.

Figure 5.1 shows a floater which consists of several separate modules. The position and orientation of module number i in the Earth-fixed frame are defined by the vector

$$\eta_i = [\eta_{i,1}, \eta_{i,2}]^T = [x_i \, y_i \, z_i \, \phi_i \, \theta_i \, \psi_i]^T. \tag{5.1}$$

where the first three degrees-of-freedom are the translations and the last three are the rotations of a given reference point at the specific module which is being considered.

The body-fixed translational and rotational *velocities* are defined by the vector

$$v_i = [v_{i,1}, v_{i,2}]^T = [u_i \, v_i \, w_i \, p_i \, q_i \, r_i]^T. \tag{5.2}$$

The body-fixed velocities are transformed to the Earth-fixed frame by application of the rotations in Eq. (5.1):

$$\dot{\eta}_i = \begin{bmatrix} \mathbf{J}_1(\eta_{i,2}) & 0 \\ 0 & \mathbf{J}_2(\eta_{i,2}) \end{bmatrix}, \tag{5.3}$$

Fig. 5.1 Earth-fixed XYZ and body-fixed xyz reference frames (From [8])

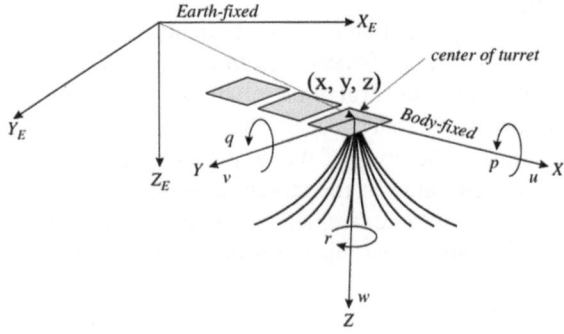

where the rotation matrices $\mathbf{J}_1(\eta_{i,2})$ and $\mathbf{J}_2(\eta_{i,2})$ are as given in [1]:

$$
\begin{aligned}
\mathbf{J}_1(\eta_{i,2}) &= \begin{bmatrix} c\psi c\theta & -s\psi c\phi + c\psi s\theta s\phi & s\psi s\phi + c\psi c\phi s\theta \\ s\psi c\theta & c\psi c\phi + s\phi s\theta s\psi & -c\psi s\phi + s\theta s\psi c\phi \\ -s\theta & c\theta s\phi & c\theta c\phi \end{bmatrix} \\
\mathbf{J}_2(\eta_{i,2}) &= \begin{bmatrix} 1 & s\phi t\theta & c\phi t\theta \\ 0 & c\phi & -s\phi \\ 0 & s\phi/c\theta & c\phi/c\theta \end{bmatrix}
\end{aligned}
\tag{5.4}
$$

where s, c and t are abbreviations for the sine, cosine and tangent functions. In the following these matrices are abbreviated as $\mathbf{J}_1$ and $\mathbf{J}_2$, respectively.

The motion of module number i in 6 degrees of freedom, is then be described by the following equation, [2]:

$$
\mathbf{M}_i \dot{v} + \mathbf{C}_i(v_i)v_i + \mathbf{D}_i(v_i)v_i + \mathbf{g}_i(\eta_i) = \tau_{i,M} + \tau_{i,E} + \tau_{i,C} + \tau_{i,T},
\tag{5.5}
$$

where $\mathbf{M}_i$ is the inertia matrix also including hydrodynamic added inertia, $\mathbf{C}_i(v_i)$ is the Coriolis and centripetal matrix, $\mathbf{D}_i(v_i)$ is the damping matrix, and $\mathbf{g}_i(\eta_i)$ represents the generalized restoring forces; $\tau_{i,M}$ corresponds to the generalized mooring forces; $\tau_{i,E}$ contains the generalized environmental forces due to wind, current and waves; $\tau_{i,C}$ corresponds to the generalized connector forces from the neighboring modules and $\tau_{i,T}$ is the vector of propulsion forces and torques. Further details of the mathematical models are found in Refs. [1–5].

Mathematical modelling within the area of cybernetics is frequently divided into two regimes, i.e. the *Process Plant Model* and the *Control Plant Model*. The former should serve as the "real world" in computer simulations, while the latter serves as a tool for design of the controller. The Control Plant Model is usually a simplified version of Process Plant Model with respect to complexity, nonlinearity and number of degrees-of-freedom. The Control Plant Model for the present case is given by:

$$
\begin{aligned}
\mathbf{M}\dot{v} + \mathbf{D}v + \mathbf{g}(\eta) &= \tau + \mathbf{J}^{\mathrm{T}}(\psi)\mathbf{b}, \\
\dot{\mathbf{p}} &= \mathbf{J}_{2\mathbf{R}}(\psi)\mathbf{w}, \\
\dot{\psi} &= \rho,
\end{aligned}
\tag{5.6}
$$

where $\boldsymbol{\eta} = [\mathbf{p}^{\mathrm{T}}, \psi]^{\mathrm{T}} = [x, y, \psi]^{\mathrm{T}}$ is the (x,y)-position and the heading in earth-fixed coordinates; the vector $v = [\mathbf{w}^{\mathrm{T}}, \rho]^{\mathrm{T}} = [u, v, \rho]^{\mathrm{T}}$ contains the translational and rotational velocities in body-fixed coordinates; $\mathbf{g}(\eta)$ represents the mooring and restoring forces, $\tau = [\tau_{\mathbf{w}}^{\mathrm{T}}, \tau_\rho]^{\mathrm{T}}$ represents the control input, $\mathbf{b} \in \mathbf{R}^3$ is a slowly varying bias term representing external forces due to wind, currents, and waves, $\mathbf{M} \in \mathbf{R}^{3\times3}$ is the inertia matrix, $\mathbf{D} \in \mathbf{R}^{3\times3}$ is the damping matrix, and $\mathbf{J}(\psi) \in \mathbf{R}^{3\times3}$ and $\mathbf{J}_{2\mathbf{R}}(\psi) \in \mathbf{R}^{2 \times 2}$ are rotation matrices defined as

$$
\mathbf{J}(\psi) = \begin{bmatrix} \mathbf{J}_{2\mathbf{R}} & \mathbf{0} \\ \mathbf{0} & 1 \end{bmatrix} = \begin{bmatrix} \cos\psi & -\sin\psi & 0 \\ \sin\psi & \cos\psi & 0 \\ 0 & 0 & 1 \end{bmatrix}
\tag{5.7}
$$

Fig. 5.2 Four connectors
between two floating modules
(From [8])

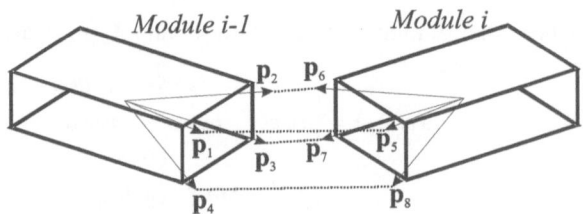

A finite element model of the mooring system is applied which is based on the approach described by Aamo and Fossen [3].

The modules in the first example below are inter-connected by rigid rod connectors which are hinged at the ends, see Fig. 5.2. The connector will impose forces on the modules as well as accompanying torques. The forces will consist of a restoring part and a viscous damping part. The former depends on the relative distance between the two modules and the latter depends on the relative velocities between them.

5.2.2 PID Control Algorithm for Dynamic Positioning

Both linear and nonlinear controllers are proposed in the literature. In [6] a linear LQG based controller is proposed, while in [7] a nonlinear PID controller is used. These controllers have been successfully installed on several commercial DP systems. In the present examples both types are applied. Since the LQG algorithm was considered in some detail in the previous chapter it is here focused on the second category.

The basic principle of a PID control law is to generate a thrust for which the different terms depend on the time-varying deviation between given target characteristics and the observed ones. The first control term is proportional to the 3-dimensional position and heading deviation vector e as referred to the vessel position relative to the desired path (which can also be a fixed position). This corresponds to the so-called proportional term. The second term is a function of the velocity deviation vector $\dot{e}$ (the differential term). The third term depends on the accumulated deviation vector (the integral term). All these vectors are referred to a specific time instant t. Based on this principle, the required thruster force vector τ_{thr} in the body-fixed frame can be formulated as:

$$\tau_{\text{thr}} = -J_e^T K_p e - K_d \bar{v} - J^T K_i \int_0^t e(\tau) d\tau \tag{5.8}$$

Here, the following quantities are employed:

$$\begin{aligned} e &= J_d^T(\eta - \eta_d) \\ \bar{v} &= v - J_d^T \dot{\eta}_d \end{aligned} \tag{5.9}$$

where J, J_d and J_e are transformation matrices. The 3-dimensional vector η_d defines the desired earth-fixed position and heading coordinates. K_p, K_d and K_i are the 3×3 non-negative controller gain matrices. η and v are the actual earth-fixed position and velocity (body-fixed) vectors of the vessel. It should be noticed that all the states are assumed to be available (which is not true in general, but state estimates based on a given observer can still be calculated).

5.3 Example of Control Scheme Based on Structural Reliability Monitoring

5.3.1 General

In the following, focus is on dynamic structural response and control schemes which apply to a sequence of stationary conditions as outlined in Chap. 3. The external excitation is assumed to contain a slowly varying load component which is due to low-frequency wind and wave forces. In the following, a "monitoring" type of control scheme which make explicit use of structural reliability criteria is first considered. The static offset position of the floating vessel is applied as a basic monitoring parameter.

5.3.2 System Model and Behavior Without Position Control

The present example is based on [8], where a model of a futuristic fish farming structure is considered. A control system is designed that: (1) ensures limited loading of the mooring system; (2) keeps the chain of surface modules aligned transversely to the incoming current, and; (3) ensures positive strain in the connectors between the modules. The control actuation is achieved by means of a thruster mounted on the first module, and a hydrofoil mounted on the last module. The performance of the control system is demonstrated by simulations, and evaluated by a structural reliability criterion based on the delta index.

The example studied is sketched in Fig. 5.3. The structure consists of five interconnected surface modules, with the first module moored to the seabed via four mooring cables. The four cables are connected to the same point on the module, allowing the surface structure to rotate freely. This configuration is motivated by several considerations. A single point mooring system, as opposed to multiple point mooring systems, is preferable from the point of view of: (1) cost effectiveness; (2) applicability in terms of the size and configuration of the attached surface structure; (3) ease of operations like attachment and detachment of surface structure; (4) modularity in terms of adding or removing individual modules, and; (5) rotational mobility of the structure, which enables continuous supply of clean

Fig. 5.3 Configuration of the marine structure consisting of five surface modules and four mooring cables (From [8])

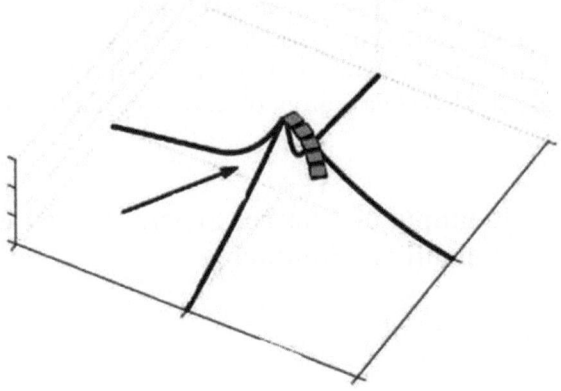

water to the fish by ensuring that fish contained in one part of the structure do not spend long periods of time in the wake of other parts of the structure

Present marine fish farms are mainly located at sheltered sites along the coast or within fjord systems. Still, some of these plants experience failure e.g. as a result of severe conditions related to wave, wind and current intensity. There are multiple causes of such failures. Some of these are related to inadequate design such as under-prediction of environmental forces or over-prediction of the real strength properties.

In the future, it is anticipated that marine fish farms will be installed and operated in more harsh environments. Even more focus will hence be on proper design criteria and procedures for approval of adequate mechanical strength properties. Furthermore, new types of fish farm structures will need to be developed for the most exposed locations. Improved robustness of these structures needs to be achieved, e.g. by means of dynamic positioning. In the following, an example of such a futuristic fish farm structure is considered. The effect of applying a thruster system is investigated.

The focus here is on problems related to tidal currents of high eccentricity together with first order wave induced motions. The surface current is assumed to be given by

$$V_c = \begin{bmatrix} v_x \\ v_y \end{bmatrix} = \begin{bmatrix} A_x \sin\left(2\pi \frac{t}{T}\right) \\ A_y \sin\left(2\pi \frac{t}{T}\right) \end{bmatrix}, \tag{5.10}$$

where $T = 12 \times 3{,}600$ (corresponding to a tidal cycle of 12 h), $A_x = 0.3$ and $A_y = 0.01$ are the amplitudes of the current in the Earth-fixed X- and Y-directions, respectively. Notice that $A_x \gg A_y$, giving a tidal ellipse with large eccentricity.

In addition to forces due to current, first order wave-induced motions are also included in the numerical simulation. This motion is found based on the vessel transfer functions. A "continuous" sea state is considered which is characterized by a significant wave height of $H_s = 2$ m, a peak period of $T_p = 8.7$ s and a mean wave direction of 60°. The simulation results are obtained with 200 wave components.

Two different versions of the structure are analysed for the purpose of comparison:

1. The fish farm is not equipped with any station-keeping facility except the mooring system.
2. The fish farm is equipped with thrusters that can be activated on command (in addition to the mooring system).

For the latter case, a monitoring scheme based on structural reliability criteria for activation of the thrusters is considered.

After an initial transient due to initial conditions, the structure enters a periodic orbit driven by the tidal current, with a period of twelve hours. Figure 5.4 shows the orbits of the five surface modules for the case *without* thrusters (i.e. the open-loop case). These modules rotate counter-clockwise. As expected, the chain of modules spends a large fraction of the time in almost a straight line, aligned with the strong current in the x-direction. This is an undesirable configuration, because modules towards the end of the chain lie in the wake of the foremost modules for long periods of time, resulting in poor environmental conditions for the fish. The high eccentricity of the tidal current also leads to an uneven loading of the mooring system. Hence, it is very relevant to apply a mechanism for motion control of the system.

The safety margin with respect to line failure can be quantified by means of an "instantaneous line index". This quantity is here defined as follows:

$$L_\delta(t) = \big(T_{Br,mean} - T_{line}(t)\big)/\sigma_{Break} \qquad (5.11)$$

where $T_{line}(t)$ is the instantaneous time-varying tension in a specific line; σ_{Break} is the standard deviation of the breaking strength of the line, which is expressed in percent of the mean value of the strength. It should be noted that the "safety measure"

Fig. 5.4 Motion of surface vessels during one tidal cycle. Open loop (From [8])

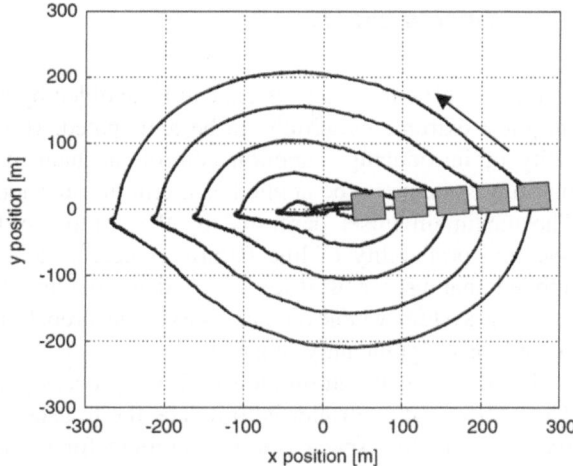

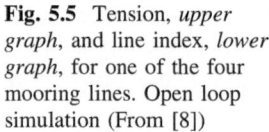

Fig. 5.5 Tension, *upper graph*, and line index, *lower graph*, for one of the four mooring lines. Open loop simulation (From [8])

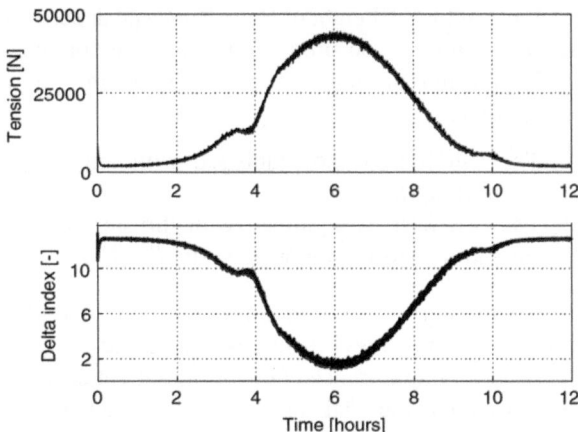

in Eq. (5.9) reflects the instantaneous behavior and hence (contrary to the delta-index defined in the previous chapter) it does not account for the possibility of a higher dynamic tension due to wave forces in the near future (i.e. within the same sea-state).

The upper graph of Fig. 5.5 shows the mooring forces in one of the mooring lines. The maximum tension is seen to be about 44 kN. For all the lines of the present mooring system, the mean breaking strength is $T_{Br;mean} = 49$ kN, and the standard deviation of the breaking strength, σ_{Break}, is taken as 7.5 % of the mean value. If the failure probability is required to be less than 10^{-3}, the corresponding lower permissible bound on the instantaneous delta index is $L_\delta(t)_{,set} = 3.1$. The time variation of the delta index is shown in the lower graph of Fig. 5.5. The minimum value of the index is seen to be about 1.5, which is well below $L_\delta(t)_{,set} = 3.1$ and therefore represents an unacceptably high failure probability.

5.3.3 System Model and Behavior with Position Control Implemented

Overload of the mooring system can be avoided by simply resizing it to withstand the most extreme conditions to be anticipated. However, by utilizing the possibility of introducing a motion control mechanism, the strength reserve of the mooring system can be applied as a criterion for activation of the control energy. The maximum offset permitted by the control system will be decisive for the resulting probability of line failure. Hence, an automatic control system is next applied that uses a thruster attached to the first floating module to assist the mooring system whenever necessary. The probability of mooring line failure is accordingly significantly reduced.

The design is based on the simplified process plant model as discussed above. The goal of the controller is to reduce the translational motions in the x-y-plane. By assuming low speed (i.e. no damping forces) and no coupling between the translational and rotational degrees of freedom, the control plant model is now expressed by (presently disregarding the internal connector forces)

$$m\begin{bmatrix} \ddot{x} \\ \ddot{y} \end{bmatrix} + \{k_0 + k(r)\}\begin{bmatrix} x \\ y \end{bmatrix} = \begin{bmatrix} l_x \\ l_y \end{bmatrix} + \begin{bmatrix} \tau_x \\ \tau_y \end{bmatrix}, \tag{5.12}$$

where m is the rigid body mass; k(r) (where $r = \sqrt{x^2 + y^2}$) is a continuously differentiable and increasing function describing the nonlinear part of the spring coefficient which represents the mooring system. Furthermore, $k(0) = \alpha$ where α is a positive integer constituting the pretension of the mooring system. l_x and l_y are the slowly varying environmental loads and τ_x, τ_y are the control inputs. Describing the spring coefficient by a nonlinear function is found to result in more accurate estimates of the missing states (as compared to a linear function), hence reducing the required actuator force.

Defining the state vector as

$$\begin{aligned} \mathbf{x}^T &= \begin{bmatrix} x_1 & x_2 & x_3 & x_4 & x_5 & x_6 \end{bmatrix} \\ &= \begin{bmatrix} x & \dot{x} & y & \dot{y} & l_x & l_y \end{bmatrix}, \end{aligned} \tag{5.13}$$

and modeling the slowly varying environmental loads as constant for a small time increment, (5.12) can be written in the standard form

$$\dot{\mathbf{x}} = \mathbf{A}\mathbf{x} + \mathbf{B}\mathbf{u} = \begin{bmatrix} \bar{\mathbf{A}} & \begin{matrix} 0 & 0 \\ 1 & 0 \\ 0 & 0 \\ 0 & 1 \\ 0 & 0 \\ 0 & 0 \end{matrix} \\ 0 \end{bmatrix} \mathbf{x} + \begin{bmatrix} \bar{\mathbf{B}} \\ \mathbf{0} \end{bmatrix} \mathbf{u}, \tag{5.14}$$

where

$$\bar{\mathbf{A}} = \begin{bmatrix} 0 & 1 & 0 & 0 \\ -\frac{k(r)}{m} & 0 & 0 & 0 \\ 0 & 0 & 0 & 1 \\ 0 & 0 & -\frac{k(r)}{m} & 0 \end{bmatrix}, \quad \bar{\mathbf{B}} = \begin{bmatrix} 0 & 0 \\ 1 & 0 \\ 0 & 0 \\ 0 & 1 \end{bmatrix}, \quad \mathbf{u} = \begin{bmatrix} \tau_x \\ \tau_y \end{bmatrix}. \tag{5.15}$$

The output $\mathbf{y}$ from the process plant is solely the position vector

$$\mathbf{y} = \mathbf{C}\mathbf{x} = \begin{bmatrix} \bar{\mathbf{C}} & \mathbf{0} \end{bmatrix} \begin{bmatrix} \bar{\mathbf{x}} \\ l_x \\ l_y \end{bmatrix} \quad \text{where } \bar{\mathbf{C}} = \begin{bmatrix} 1 & 0 & 0 & 0 \\ 0 & 0 & 1 & 0 \end{bmatrix} \tag{5.16}$$

Having applied a linearizing feedback, the separation principle of linear systems can be utilized. In order to solve the state feedback problem, the environmental forces can be cancelled by setting

$$\mathbf{u} = -\begin{bmatrix} l_x \\ l_y \end{bmatrix} + \mathbf{v} \tag{5.17}$$

in order to obtain

$$\dot{\bar{x}} = \bar{A}\bar{x} + \bar{B}u \tag{5.18}$$

A state feedback control law can then be designed by application of LQG procedures on the following form:

$$v = \bar{K}\bar{x}, \tag{5.19}$$

where v contains the "surplus thruster forces", which achieves desired rates of convergence of the position vector to the origin.

Driving the states to the origin by the thruster force would make the mooring system superfluous. The objective of the thruster is instead to assist the mooring system in the event that the center of the turret moves so far away from the origin, (due to extreme environmental loads), that a significant risk of line failure arises. Thus, we define r_m to be the maximum distance the center of the turret should move. r_m defines in turn the maximum total energy the system is allowed to have, $E_m = \frac{1}{2}k_0(x^2 + y^2)$, which is the potential energy stored in the mooring system when $x^2 + y^2 = r_m^2$. The total energy of the system is written as

$$E(\bar{x}) = \frac{1}{2}m(\dot{x}^2 + \dot{y}^2) + \frac{1}{2}k_0(x^2 + y^2) + \int_0^{\sqrt{x^2+y^2}} k(r)dr \tag{5.20}$$

In order to obtain a continuous (nonlinear) control, a lower energy threshold is defined, $E_t < E_m$, at which point the control action is gradually activated according to

$$v = f(E)\bar{K}\bar{x}, \tag{5.21}$$

where

$$f(E) = \begin{cases} 0, & E \leq E_t \\ \frac{E^2}{\Delta E^2} - 2\frac{E_t}{\Delta E^2} + \frac{E_t^2}{\Delta E^2}, & E_t < E < E_m \\ 1, & E > E_m \end{cases} \tag{5.22}$$

$$\Delta E = E_m - E_t.$$

The upper and lower threshold values are defined in terms of response energy. By assuming that the kinetic energy is identically zero, the corresponding threshold values for the vessel offset can be computed. However, since a small percentage of the total energy will be contributed by kinetic energy, the "true" offset values will be somewhat lower than these values.

Criteria are required for selection of lower and upper bounds for the function f(E) which are based on consideration of strength parameters for the mooring lines. These are obtained by application of the delta index which was introduced in (5.9), evaluated as a function of the offset of the foremost vessel. For given values of this index, the corresponding probability for the event that failure of the mooring line with the highest tension may occur can be estimated. For the present mooring system, it is estimated that the failure probability for a maximum offset value of 55 m is equal to 10^{-3} per year (based on the same statistical characteristics as applied above for both the breaking strength and the extreme response during the annual largest sea state. This corresponds to a delta index of 2.9). If the corresponding lower threshold value for the offset is set to 52.3 m, the resulting annual failure probability for that (mean) offset is estimated as 10^{-4} (corresponding to a delta index of 3.5).

The target value for the failure probabilities (in particular the probability corresponding to the critical maximum offset) should be selected by consideration of costs and consequences associated with failure of a mooring line. The energy consumption which depends on how often the thruster forces are activated (i.e. the value of the lower threshold for the offset) will also enter the picture.

The present procedure for calculation of the failure probability represents a very simplified approach. In reality, the failure probability should be evaluated based on the following expression:

$$P(T_{\max} > T_{break}), \tag{5.23}$$

where $T_{\max} = \max(T_{Slow}(t) + T_{Dyn}(t))$, $T_{Slow}(t)$ is the slowly time-varying mean tension and $T_{Dyn}(t)$ is the dynamic tension induced by time-varying actions excluding the current. However, the present purpose is to illustrate the effect of strength criteria on the control algorithm, and the simplified scheme based on application of the delta index is accordingly adequate.

The only likely measurement to be available is position, provided by a GPS receiver. Therefore, the controller suggested in the previous section should not be implemented directly. However, estimates of the state vector can be applied as a replacement for the "true" and unknown vector. An observer that provides an estimate of the states can be constructed by adding an output injection term, giving

$$\dot{\hat{\mathbf{x}}} = \mathbf{A}\hat{\mathbf{x}} + F(\mathbf{x}) + \mathbf{B}\mathbf{u} + \mathbf{L}(\mathbf{y} - \mathbf{C}\hat{\mathbf{x}}). \tag{5.24}$$

Notice that $F(\mathbf{x})$ contains the nonlinear term of the spring force introduced by the mooring system. It is a function of measurable states x_1 and x_3. Inclusion of nonlinear effects for the mooring system was found to improve the estimates of the environmental loads significantly.

The matrix $\mathbf{L}$ can be designed to achieve any desired rate of convergence of the estimated state vector $\hat{\mathbf{x}}$ to the true one, $\mathbf{x}$. A LQG design is carried out to find the tuning parameters. These parameters have been chosen such that the observer gives satisfactory wave-filtering properties.

In summary, the thruster assistance control system is a nonlinear dynamic output feedback controller, given by

$$\begin{aligned} \dot{\hat{\mathbf{x}}} &= \tilde{\mathbf{A}}(\hat{E})\hat{\mathbf{x}} + \mathbf{L}\mathbf{y}, \\ \boldsymbol{\tau} &= \mathbf{K}(\hat{E})\hat{\mathbf{x}} + k(r)\mathbf{y}. \end{aligned} \tag{5.25}$$

where

$$\hat{E} = \frac{1}{2}m(\hat{x}_2^2 + \hat{x}_4^2) + \frac{1}{2}k_0 \cdot \mathbf{y}^T\mathbf{y} + \int\limits_0^{\sqrt{\mathbf{y}^T\mathbf{y}}} k(r)rdr$$

$$\mathbf{K}(\hat{E}) = \begin{bmatrix} f\hat{E}\bar{\mathbf{K}} & 0 \\ 0 & -\mathbf{I} \end{bmatrix} \tag{5.26}$$

$$\tilde{\mathbf{A}}(\hat{E}) = \mathbf{A} + \mathbf{B}\mathbf{K}(\hat{E}) - \mathbf{L}\mathbf{C} \tag{5.27}$$

Exponential attractiveness of the region defined by $E \le E_{max}$ is claimed by the separation principle of linear time invariant systems. It has been found that using a nonlinear function describing the stiffness properties of the mooring system (instead of a linear relationship) has significantly improved the estimates of the environmental loads, see [8] for further details.

A closed loop simulation is performed based on specified upper and lower values for the threshold energies which define the function f(E). The offset threshold values are selected by application of the basic version of the delta index as discussed above. As mentioned, the offset threshold values for the two considered failure probability levels are equal to 52.3 and 55 m, respectively.

The effect of including nonlinear terms for describing the stiffness properties of the mooring system has been studied by comparing results for the same scenarios with a linearized stiffness. An average decrease of 5 % of the power consumption was observed. An even more important effect of the nonlinear observer is the ability to estimate the environmental forces accurately.

For a preset offset limit of 55 m, the maximum observed value of r is approximately 53.5 m which is somewhat below the requirement. The maximum r is hence reduced only slightly as compared to the open-loop case. The modified variation of the surface vessel position during one tidal cycle is shown in Fig. 5.6. The corresponding tension and delta index of the most loaded mooring line, compared to the open loop results, are shown in the upper and middle graph of Fig. 5.7. The maximum value for the critical line is now around 37 kN, with the corresponding value of the delta index being close to δ_{set}. This is achieved by using the thrusters approximately 35 % of the time (Fig. 5.7, lowermost part).

With reference to the present example, a model of a futuristic fish farming structure has been developed and is studied with respect to strategies for configuration control. For a chain of modules moored to the seabed a thruster-based control system is designed that: (1) ensures limited loading of the mooring system in order to avoid cable breakage; (2) keeps the modules aligned transversely to the incoming current in order to ensure continuous supply of clean water to the fish, and; (3) implies positive strain in the connectors between modules in order to avoid buckling effects in turning currents, such as tidal currents with high

Fig. 5.6 Position of surface vessels during one tidal cycle with thruster position control (From [8])

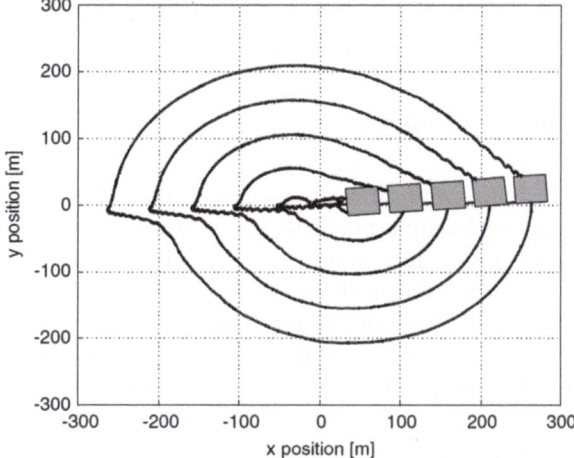

Fig. 5.7 Tension in critical mooring line, *upper graph*, instantaneous delta index for the critical mooring line, *middle graph*, and thruster output, *lower graph* (From [8])

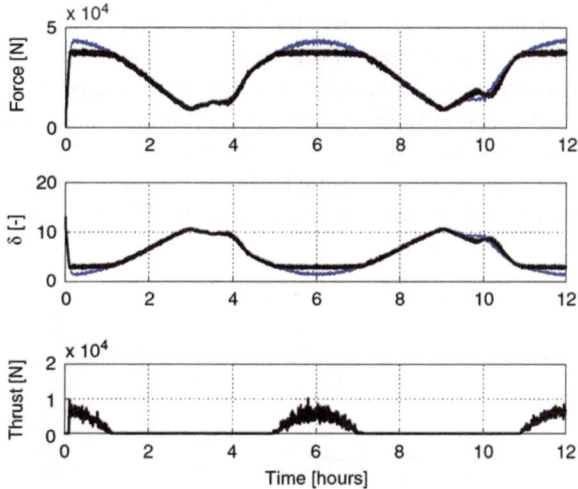

eccentricity. Actuation was done by means of a thruster mounted on the first module, and a hydrofoil mounted on the last module. Future work might involve active connectors between surface modules for motion damping, and in particular damping of wave-induced motions. Another possible development is the inclusion of the reliability index as a continuous and intrinsic part of the control law.

5.4 Control Schemes Based on On-Line Computation of Reliability Measures

5.4.1 Example 1

The following example is mainly based on [9–12] where dynamic positioning of a semi-submersible platform in relation to control of riser angles are considered. A schematic illustration of the vessel and the riser system is shown in Fig. 5.8.

Several problems have been experienced during marine drilling operations due to excessive top and bottom riser angle response levels. For the upper part of the riser, contact between the riser pipe and the surface vessel may easily lead to serious damage. For the lower part, even moderate angles (2–4°) may imply that the drill-pipe within the riser gets into contact with the ball-joint or the well-head. Wear due to metal-to-metal contact implies that damage of the well-head may occur over time, and in some cases a blow-out at the seabed can be the final result. For large riser angles (>4–6°) at the seabed, the operation has to be interrupted and for increasingly larger angles (>6–7°) a controlled disconnect of the lower part of the riser is required. The actual limits for the riser angle depend on the type of riser and the blow-out-preventer (BOP) which is applied, in addition to the type of subsea installation. If the bottom angle increases too quickly, an emergency disconnect is activated automatically on many installations.

It is accordingly of interest to minimize the response levels for the angles. One way of achieving this is by moving the surface floater to a proper position. If mooring lines are applied, this is not a continuous process but will rather be performed at selected time instants. If a dynamic positioning (DP) system is

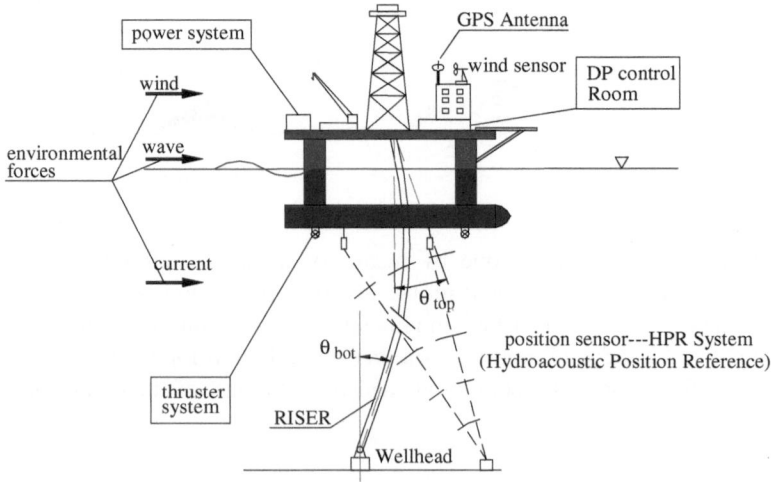

Fig. 5.8 General view of the riser-floater system (From [12])

applied, the attractive option arises to implement riser response criteria within the position control loop. However, control of the wave-frequency motions of the vessel is both unrealistic and generally unnecessary. Instead, it is typically aimed at controlling the slowly varying low-frequency (LF) motions from wind loads, second order and mean wave loads and time-varying current loads.

A basic problem is that minimizing the response level for one of the angles will typically imply an increase of the response level for the other angle (somewhat depending on the variation of the current profile as a function of depth). Accordingly, relative weights must be put on the criteria for the top versus the bottom angle. An attractive approach is to express these weights as functions of the respective reliability indices for each of the two angles. A further possibility is to apply an objective function which is purely expressed in terms of reliability indices. The viability of different schemes of this type is explored by numerical simulation for a specific riser which operates at a water depth of 1,000 m.

Applications of the present control scheme to other related areas would comprise cases where the quantity to be controlled is a random process with slowly developing "global characteristics" such as mean value and variance. The short-term behavior of such a process is then characterized by a constant mean value with random dynamic fluctuations around this value.

The optimal position of the vessel, η_{opt}, can be computed by taking the riser angle criteria into account, based on the measured values of the angles at any time. However, the vessel cannot be moved to the specified optimal position instantaneously. Instead, a smooth transition is required. Therefore, the transition path for the position and heading η_d can be obtained if such a smooth reference model e.g. is introduced. In order to provide high-performance of the DP vessel's operations, a third-order reference model is usually chosen, [6].

If the riser angles are not considered, the optimal position of the vessel is typically taken to be just above the well-head. This will generally not be optimal if criteria related to the angles are included. A better alternative can be achieved by introducing a quadratic objective function based on the top and bottom angles, which is of the type:

$$L(\alpha_t, \alpha_b) = [w_t(\alpha_{tx}^2 + \alpha_{ty}^2) + w_b(\alpha_{bx}^2 + \alpha_{by}^2)] \tag{5.28}$$

where $(\alpha_{tx}, \alpha_{ty})$ are the x- and y-components of the top angle, respectively, and $(\alpha_{bx}, \alpha_{by})$ are the x- and y-components of the bottom angle; w_t and w_b are the corresponding weighting factors for the respective angles.

The angular components in this expression can in turn be expressed as the sum of the instantaneous measured angle components and the incremental components due to an increment of the vessel position. The angular incremental components are in turn expressed as explicit linear functions of this incremental vessel position by means of the so-called "influence coefficients". These four influence coefficients (c_{tx}, c_{ty}) and (c_{bx}, c_{by}) are basically obtained from a numerical model of the riser e.g. by application of the Finite Element Method.

These coefficients represent the change of each angle component given a unit change of the vessel position. In principle, these coefficients will change as functions of the vessel position due to the nonlinear geometric behavior of the riser. They will also change if the top tension of the riser changes. Furthermore, they can be anticipated to change as functions of the surface current velocity and the current profile (i.e. the velocity variation as a function of depth).

Accordingly, these coefficients should be calculated at each time step based on a pre-established riser model which is subjected to the proper static loads at that step. However, significant savings in computational effort can be achieved if these coefficients are established in advance. This possibility obviously depends on the stability of the coefficients for varying vessel offset and varying current profiles (assuming that the riser weight and top tension is constant). This topic is addressed below.

Having established the influence coefficients, the optimal magnitude of the increment of the vessel position and the optimal direction of this increment can be found by differentiating the objective function. By setting the derivatives with respect to the position increments in the x- and y-directions equal to zero, the minimum value of the object function is identified. The optimal increment of vessel position is found to be:

$$\Delta r^*_{vessel} = \left(w_t c_{tx} \alpha_{tx} \cos \theta_{opt} + w_b c_{bx} \alpha_{bx} \cos \theta_{opt} + w_t c_{ty} \alpha_{ty} \sin \theta_{opt} + w_b c_{by} \alpha_{by} \sin \theta_{opt} \right)$$
$$/ \left(w_t c_{tx}^2 (\cos^2 \theta_{opt}) + w_b c_{bx}^2 (\cos^2 \theta_{opt}) + w_t c_{ty}^2 (\sin^2 \theta_{opt}) + w_b c_{by}^2 (\sin^2 \theta_{opt}) \right)$$

(5.29)

and the corresponding optimal direction is given by:

$$\theta_{opt} = \tan \left(\frac{\Delta y}{\Delta x} \right)$$

(5.30)

where

$$\Delta y = (w_b c_{bx}^2 + w_t c_{tx}^2) \cdot (w_b c_{by} \alpha_{by} + w_t c_{ty} \alpha_{ty})$$

(5.31)

$$\Delta x = (w_b c_{by}^2 + w_t c_{ty}^2) \cdot (w_b c_{bx} \alpha_{bx} + w_t c_{tx} \alpha_{tx})$$

(5.32)

Here, $(\alpha_{tx}, \alpha_{ty})$ are the x-and y-components of the measured top angle. The corresponding components for the measured bottom angle are designated as $(\alpha_{bx}, \alpha_{by})$. The optimal vessel position set-point is then obtained as:

$$\eta_r^* = \eta_r + \Delta r^*_{vessel} \left[\cos \theta_{opt} \ \sin \theta_{opt} \ 0 \right]^T$$

(5.33)

It is of some interest to consider the special case of two-dimensional loading and motion within the x-z plane only. The optimal direction is then along the x-axis, and the optimal position increment becomes:

$$\Delta r^*_{vessel} = (w_t c_{tx} \alpha_{tx} + w_b c_{bx} \alpha_{bx}) / \left(w_t c_{tx}^2 + w_b c_{bx}^2 \right)$$

(5.34)

which is of a much simpler form than for the three-dimensional case. A further simplification is obtained if only a single riser angle is to be controlled (e.g. the bottom angle). The result then reads:

$$\Delta r^*_{vessel} = \alpha_{bx}/c_{bx} \tag{5.35}$$

which is simply the instantaneous angle divided by the influence coefficient for the bottom angle.

As discussed above, the "influence coefficients" for the riser angle components play a key role in the dynamic positioning based on optimal set-point chasing. These coefficients are discussed in the following with the aim of obtaining simplified relationships.

In order to limit computation time, it is of interest to minimize the number of finite elements which is applied for the riser model, while still maintaining a sufficient level of accuracy. It was found in Chen [12] that for around 10 elements, the error as compared to a very fine discretization was less then 10 % for the particular deep water riser which was considered. Increasing the number of elements to 50, the error was reduced to less than 1 %. For practical purposes, it seems that between 10 and 20 elements will be sufficient. It is anticipated that for this particular type of response, the selected number of elements is indicative for what is to be expected also for other riser configurations of the top-tensioned type.

Linearity of the relations between the angles and the vessel offset was also confirmed for a number of 2D current profiles in [12]. Turning to 3D current profiles, the same conclusion is made with respect to the X- and Y-components of the top and bottom riser angles when these are considered separately. This is shown in Fig. 5.9 for the same riser configuration but with a rotating current profile. For this case the direction of the current at the surface is along the global X-axis, while it is rotated by 180° at the seabed.

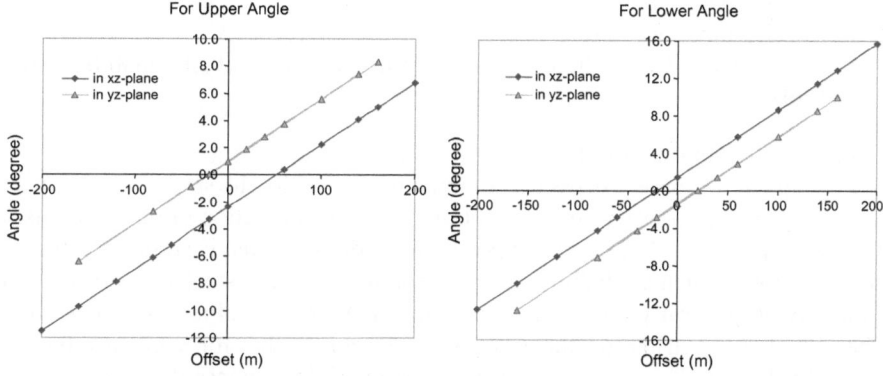

Fig. 5.9 X- and Y-components of *upper* (*left*) and *lower* (*right*) riser angle as functions of vessel offset for a 3D current profile with direction of velocity vector rotated by 180° *from top to bottom* (From [12])

The computed static response clearly suggests that linear relations provide very close approximations to the results obtained by application of a detailed finite element representation at each time step. This implies that the influence coefficients described above can be regarded as constants. This clearly also suggests that the coefficients can be computed a priori, i.e. outside the control loop.

As already discussed, the most basic dynamic positioning scheme corresponds to application of a fixed reference set-point. A second step is provided by the scheme outlined above which introduces a set-point that is continuously updated, which accounts for riser response criteria. The instantaneous values of the angles that enter the expression for the incremental offset should not include dynamic response components. This is due to the relatively rapid change of the angles within the frequency band of the wave energy (e.g. from 1 to 20 s). This applies in particular to the upper riser angle for which the dynamic response level typically exceeds the static one. Rather, smoothed values of the angle time series need to be established. The mean values of the two angles (as functions of time) immediately present themselves as good candidates. They are easy to estimate and change only slowly as functions of time.

A weakness of such an approach is that the extreme response levels are not reflected properly. This implies that for a given time period, there will be a certain probability that the maximum permissible response values are exceeded, i.e. that "failure" occurs. This is even more true for the positioning algorithm based on a fixed set-point since for that case the response mean-values are not properly adjusted to prevent failure.

Hence, a compact representation of the response process which comprises both the mean value and the variance is demanded. Such a representation is provided by the structural reliability index, which in addition accounts for the permissible response threshold in a proper way. For the present example there are at least three different alternatives for how reliability indices can enter the control loop. These are here referred to as:

1. Reliability-index monitoring.
2. Reliability-index weighting.
3. Control actions based directly on reliability indices (direct reliability-index control).

In the following the third category is considered.

For a given response process (or combinations of such), the structural beta index can be calculated for each time step. Furthermore, computation of this index itself does not depend on which dynamic positioning controller that is applied. However, the definition of such a reliability index is not unique. For a stochastic process, the reliability is generally formulated in terms of extreme values as discussed in Chap. 3. If the distribution function of the extreme value for a given reference interval is known, $F_{R_{extr}}(r_{extr})$, and a given permissible response threshold value is given, $r_{threshold}$, the corresponding reliability index can be expressed as:

$$\beta = -\Phi^{-1}(p_f) = -\Phi^{-1}(1 - F_{Rextr}(r_{threshold})) \tag{5.36}$$

where p_f refers to the probability of failure (which is equal to the complement of the cumulative extreme value distribution function); Φ^{-1} designates the inverse standard normal cumulative distribution function. This index can obviously be generalized to include the effect of additional uncertainties related e.g. to the response threshold itself. However, computation of the reliability index as presently defined involves several challenging tasks:

1. The response processes for offshore structures are generally non-stationary due to time-varying environmental characteristics. This implies that it is convenient to introduce a representative duration for which the process can be modeled as being stationary. This duration is typically taken to be of the order of an hour. The extreme value of the response will then also refer to the same duration.
2. Estimating the parameters of the extreme-value distribution is associated with inherent statistical uncertainty. Furthermore, it may also be required to identify which type of extreme-value distribution to apply for the response e.g. due to nonlinear effects.

As a consequence of these complexities, it is relevant to consider a simplified "instantaneous" version of the reliability index such as:

$$\beta_{simp} = (r_{threshold} - E[r])/\sigma_r \tag{5.37}$$

where $E[r]$ is the (estimated) mean value of the response and σ_r is the (estimated) standard deviation of the response process. This simplified index is still able to capture both the static and dynamic response components. Furthermore, it requires only a continuous estimation of the mean value and variance of each response process. An additional benefit of this index is that it can be applied more directly for derivation of dynamic positioning criteria based on objective functions that are explicitly expressed in terms of this index.

A primary candidate for such an objective function is the following:

$$L\left(\alpha_t, \alpha_b, t\right) = (\beta_{CR,t} - \beta_t(t))^2 + (\beta_{CR,b} - \beta_b(t))^2 \tag{5.38}$$

where the time dependence of the reliability indices is explicitly represented. The subscripts b and t in this equation refer respectively to bottom and top angles. The subscript CR refers to the critical value of the index.

In order to identify closed-form solutions for the minimum point of this objective function the simplified version of the reliability index is applied. Furthermore, it is assumed that the relation between the change of the mean values of the riser angles and the vessel offset is the same as for the static riser response which was discussed above. The two-dimensional version of the objective function is now considered for simplicity (and without loss of generality). If the variances of the response processes are unchanged by a given vessel position increment, the explicit form of the object function becomes:

$$L(E[\alpha_t], E[\alpha_b], \sigma_{\alpha t}, \sigma_{\alpha b}) = (\beta_{CR,t} - ((\alpha_{CR,t} - (E[\alpha_t] + c_t r))/\sigma_{\alpha t}))^2$$
$$+ (\beta_{CR,b} - ((\alpha_{CR,b} - (E[\alpha_b] + c_b r))/\sigma_{\alpha b}))^2 \tag{5.39}$$

where $\sigma_{\alpha t}$ and $\sigma_{\alpha b}$ are the standard deviations of the top and bottom angle response processes, respectively.

However, for the case that the top and bottom angles have different signs, this objective function has to be modified slightly. Assuming e.g. that the mean value of the bottom angle is negative, the second term is modified and we get:

$$L(E[\alpha_t], E[\alpha_b], \sigma_{\alpha t}, \sigma_{\alpha b}) = (\beta_{CR,t} - ((\alpha_{CR,t} - (E[\alpha_t] + c_t \Delta r))/\sigma_{\alpha t}))^2$$
$$+ (\beta_{CR,b} - ((\alpha_{CR,b} + (E[\alpha_b] + c_b \Delta r))/\sigma_{\alpha b}))^2 \tag{5.40}$$

The value of the incremental offset position which minimizes this objective function is then found as:

$$\Delta r = \{(-c_b \sigma_{\alpha t}^2 \beta_{CR,b} \sigma_{\alpha b} - c_b \sigma_{\alpha t}^2 \alpha_{CR,b} + c_b \sigma_{\alpha t}^2 E[\alpha_b] - c_t \sigma_{\alpha b}^2 \beta_{CR,t} \sigma_{\alpha t}$$
$$+ c_t \sigma_{\alpha b}^2 \alpha_{CR,t} + c_t \sigma_{\alpha b}^2 E[\alpha_t])/(c_b^2 \sigma_{\alpha t}^2 + c_t^2 \sigma_{\alpha b}^2)\} \tag{5.41}$$

It is of special interest to investigate the particular case that only one of the riser angles is to be controlled. Without loss of generality, we can assume that this is the bottom angle. The optimal repositioning of the surface vessel is then obtained by keeping only the second term of the object function, and the optimal offset increment is then expressed as:

$$\Delta r = \{(-\beta_{CR,b} \sigma_{\alpha b} - \alpha_{CR,b} + E[\alpha_b])/c_b\} \tag{5.42}$$

which is to be compared to expression (5.41) above.

The quadratic form of the objective function in Eq. (5.40) implies that the same penalty is put on too low reliability indices as on too high indices. Furthermore, if the reliability index becomes higher than the target value, energy is spent in decreasing the reliability index (i.e. making the system more unreliable). This can possibly be regarded as a weakness of the present formulation. An additional issue is the lack of symmetry of the objective function if the mean value of the response process crosses from positive to negative. For that case, the sign of the permissible threshold of the angle should also be changed which would imply a non-constant objective function.

To amend these unwanted properties, it is relevant also to consider the following objective function which is symmetric with respect to positive and negative threshold response values:

$$L(E[\alpha_t], E[\alpha_b], \sigma_{\alpha t}, \sigma_{\alpha b}) = (\beta_{CR,t} - ((\alpha_{CR,t} - (E[\alpha_t] + c_t \Delta r))/\sigma_{\alpha t}))^2$$
$$+ (\beta_{CR,b} - ((\alpha_{CR,b} - (E[\alpha_b] + c_b \Delta r))/\sigma_{\alpha b}))^2$$
$$+ (\beta_{CR,t} - ((\alpha_{CR,t} + (E[\alpha_t] + c_t \Delta r))/\sigma_{\alpha t}))^2 \tag{5.43}$$
$$+ (\beta_{CR,b} - ((\alpha_{CR,b} + (E[\alpha_b] + c_b \Delta r))/\sigma_{\alpha b}))^2$$

This objective function can be regarded as corresponding to a two-sided barrier with critical values of the top and bottom angles being located symmetrically around the static mean value (i.e. with +/− signs). For this case, the possibility that the mean values of the angles have different signs does not require any special modification.

The minimum value of this object function is now obtained for an incremental vessel offset which is expressed as:

$$\Delta r = (c_b \sigma_{\alpha t}^2 E[\alpha_b] + c_t \sigma_{\alpha b}^2 E[\alpha_t]) / (c_b^2 \sigma_{\alpha t}^2 + c_t^2 \sigma_{\alpha b}^2) \tag{5.44}$$

This formula can be rewritten on the form:

$$\Delta r = [(c_b / \sigma_{\alpha b}^2) \, E[\alpha_b] + (c_t / \sigma_{\alpha b}^2) E[\alpha_t] / (c_b^2 / \sigma_{\alpha b}^2) + (c_t^2 / \sigma_{\alpha t}^2)] \tag{5.45}$$

where we have divided by the factor $(\sigma_{\alpha Bot}^2 \cdot \sigma_{\alpha Top}^2)$ both in the numerator and in the denominator.

A striking property of this solution is that neither the threshold value of the response nor the critical reliability index enter the final expression. It is seen that the weights associated with the two angles are inversely proportional to their variances (i.e. which also reflect the relative variability and uncertainty). However, the angular coefficients c_t and c_b modify the weighting somewhat as compared to a direct "inverse variance" weighting. If these two coefficients are identical, the expression specializes to a direct variance-inverse weighting.

For the case that only a single response process (e.g. lower angle) is to be controlled, the optimal vessel increment becomes:

$$\Delta r = E[\alpha_{Bot}] / c_b \tag{5.46}$$

which coincides with the last term obtained from the "unsymmetric" objective function above.

In the numerical example below, results obtained by application of both the objective function in Eqs. (5.40) and (5.43) are compared. [Note that when performing the comparison of the different control strategies, the "correct" reliability indices which correspond to Eq. (5.36) are applied instead of the simplified ones. The latter are rather applied within the control loop in order to achieve a sufficiently robust and simple expression for the control action (i.e. as expressed in terms incremental vessel offset)].

A simulation study of a dynamically positioned semi-submersible vessel conducting offshore drilling is carried out to demonstrate the effect of introducing criteria related to the riser response. The floater is a semi-submersible which is equipped with 4 azimuthing thrusters, each of which are able to produce 1,000 kN. These are located at the four corners at the two pontoons.

The operational draught is equal to 24 m, the vessel mass corresponding to this draught is 45,000 tons, the length is 110 m, and the breadth is 75 m. The radius of gyration in roll is 30 m, in pitch it is equal to 33 m and in yaw 38 m. The undamped resonance periods in roll and pitch are found to be equal to 55 and 60 s, respectively.

Fig. 5.10 Surface current
velocity as a function of time
(From [9])

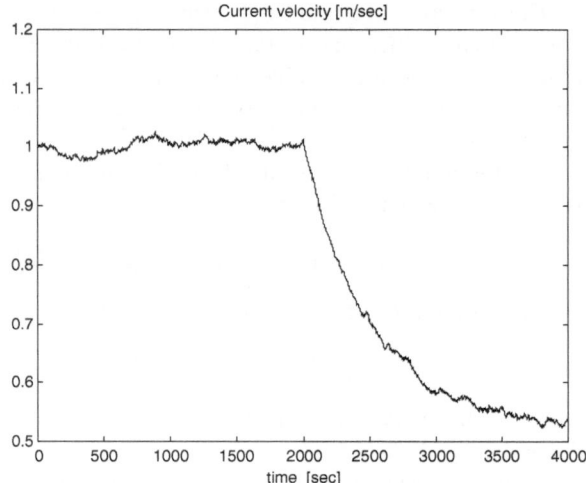

The current profile and the current surface velocity as a function of time are assumed to be known. The latter is shown in Fig. 5.10. A time window of 4,000 s is applied in the simulation. Here, the mean values and variances of the top and bottom angle response processes are assumed to be known based on measurements (However, as discussed above the estimation of these values based on measured response represents a topic on its own.).

The following variance is assumed to apply for the top angle: It is constant and equal to 0.2 (degrees)^2 until 500 s, and subsequently it increases linearly to 0.55 (degrees)^2 at 4,000 s. The variance of the bottom angle is one-fourth of the variance for the top angle. Presently the critical value for the top angle is set to 5° while it is set to 2.5° for the bottom angle.

The simplest dynamic position scheme is based on a fixed set-point which is typically just above the well-head. For this case, the reliability indices for the top and bottom angles can be computed as a function of time. The "correct" version of the reliability index is applied, which was defined in Eq. (5.36) above. The probability of failure is computed based on a Gumbel extreme value distribution as referred to a 20 min stationary period. The probability is obtained from the corresponding cumulative distribution function by inserting the critical values of the respective angles.

The mean values of the two angles as functions of time are shown in Fig. 5.11. The dynamic response components need to be added to these time series in order to get the total response. These dynamic components are clearly different for the two angles.

The corresponding reliability indices as functions of time are shown in Fig. 5.12. It is seen that the reliability index for the bottom angle becomes zero and negative in the time interval from 1,100 to 2,000 s. This implies that the failure probability exceeds 50 %. Typically, a much smaller probability would be permitted, in particular for the bottom angle. If a "reliability index monitoring" is

Fig. 5.11 Mean values of *top* and *bottom angles* for fixed set-point above well-head (From [9])

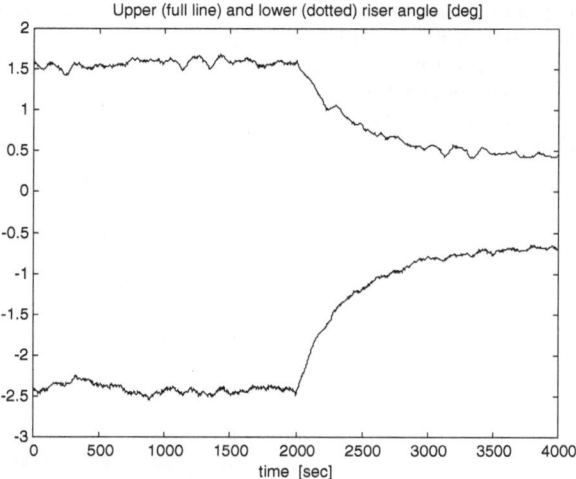

Fig. 5.12 Reliability indices as functions of time for fixed set-point above well-head (From [9])

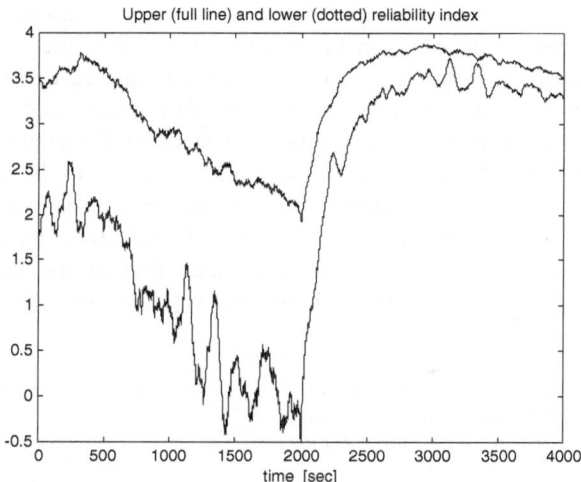

performed, some kind of action should hence be undertaken. Such an action could e.g. correspond to moving the surface floater.

The most direct way of taking the riser angle criteria into account in the dynamic positioning algorithm is by application of Eqs. (5.29)–(5.35). This algorithm is based on the instantaneous mean values of the riser angles. For the present application the weight for the bottom angle (w_b) is taken to be 5 times that for the top angle (w_t). For such a positioning algorithm, the dynamic response components are excluded from the control loop itself. However, "reliability index monitoring" (or monitoring of maximum dynamic measured angles) can still be performed to decide when counteracting measures need to be applied.

Fig. 5.13 Mean values of *top* and *bottom angles* for positioning scheme with fixed weights. Control scheme is activated at t = 600 s (From [9])

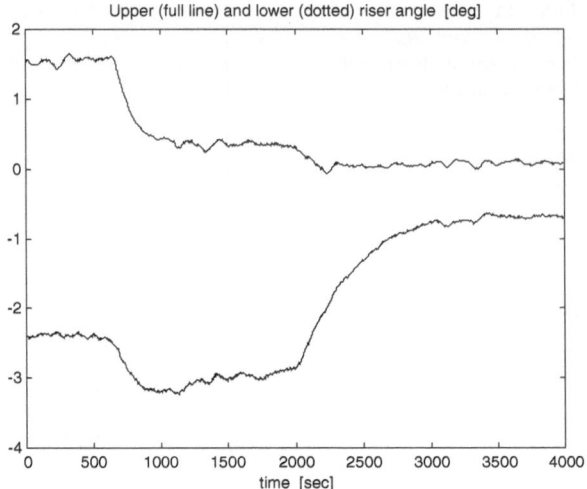

The mean values of the two angles as functions of time based on this control scheme are given in Fig. 5.13. Activation of the control scheme is performed at time t = 600 s. Prior to this instant, the surface vessel is positioned right above the well-head. When the control algorithm is activated, the mean value of the bottom angle is reduced due to the highest weighting factor for this angle. Similarly, the mean value of the top angle is increased due to the smallest weight for this angle.

The corresponding values of the reliability indices as functions of time are shown in Fig. 5.14. It is seen that the reliability index associated with the bottom angle now is very high after activation of the control algorithm. The lowest value is slightly above 4, and this value occurs at the end of the simulation period

Fig. 5.14 Reliability indices as functions of time based on the mean values in Fig. 5.13 (From [9])

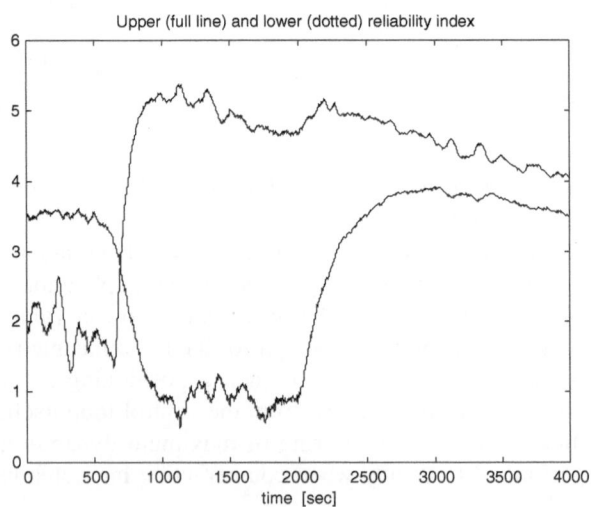

Fig. 5.15 Vessel position as function of time for positioning scheme with fixed weights ($w_b = 5$ and $w_t = 1$) (From [9])

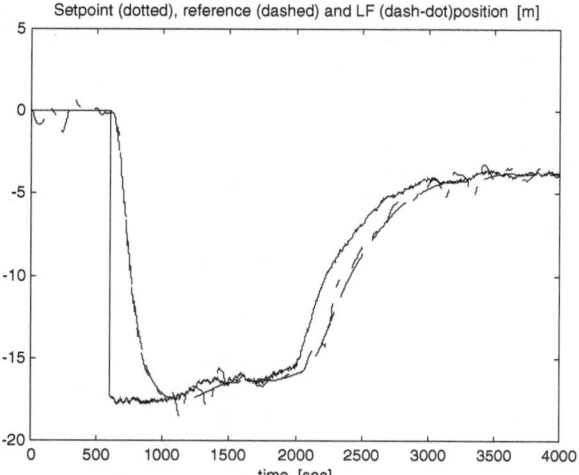

(the target value of the reliability index for the top angle is taken to be 1.0, while that for the lower angle is 2.0 i.e. a higher reliability level). The lowest value of the reliability index for the top angle is around 0.5 which is below the target value.

The surface vessel motion which results from the present positioning scheme with fixed weighting factors is shown in Fig. 5.15. When the control action is initiated, the setpoint corresponds to offset values in the interval between -15 and -20 m. When the current is reduced, the set-point moves towards zero and stabilizes around -4 m. The smoothed reference trajectory is also shown, and the vessel position is seen to oscillate around this reference path.

The positioning schemes based on a fully reliability-based objective function are next investigated. We first consider the "unsymmetric" version which corresponds to the expression given in Eq. (5.39). The variation of the mean values of the angles as function of time is shown in Fig. 5.16. The mean value of the bottom angle is reduced just enough during the critical period at the cost of a rather small increase (as compared to the previous control option) for the top angle.

The reliability indices corresponding to this strategy are shown in Fig. 5.17. It is seen that now the index for the top angle stays above 1.0, which is set as the critical value. The reliability index for the bottom angle stays above the critical value of 2.0 after activation of the control scheme. As mentioned, the present critical values of the reliability indices are formulated in terms of the "correct" index based on extreme-value statistics. However, the values of the indices which enter the control algorithm are based on the "simplified" index. Accordingly, the critical values for these indices need to be selected in a proper way in order to achieve the desired values for the "correct "indices. Based on a Gaussian response process model, the conversion between the two pairs of indices is evaluated as: (2.5, 4.2) for the "simplified" indices versus (1.0, 2.0) for the "correct" indices, where the first value in each pair refers to the top angle. This conversion will in general be influenced by the degree of non-Gaussian response behavior.

Fig. 5.16 Mean values of *top* and *bottom angles* for positioning scheme based directly on simplified reliability indices (unsymmetric objective function). *Bottom angle* $\theta_{cr} = 2.5°$, $\beta_{cr} = 1$. *Top angle* $\theta_{cr} = 5°$, $\beta_{cr} = 2.0$ (From [9])

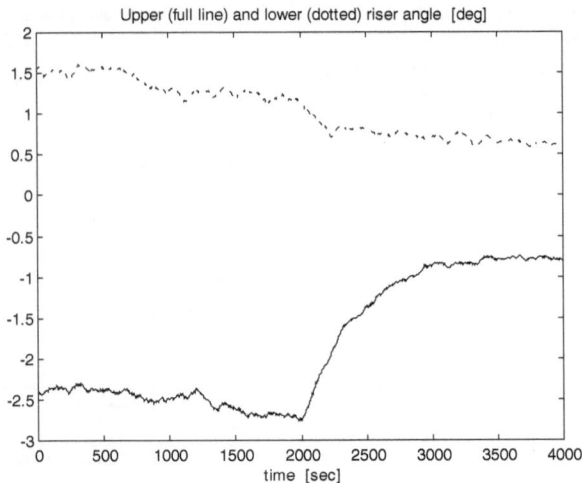

Fig. 5.17 Reliability indices as functions of time based on the mean values in Fig. 5.16 (From [9])

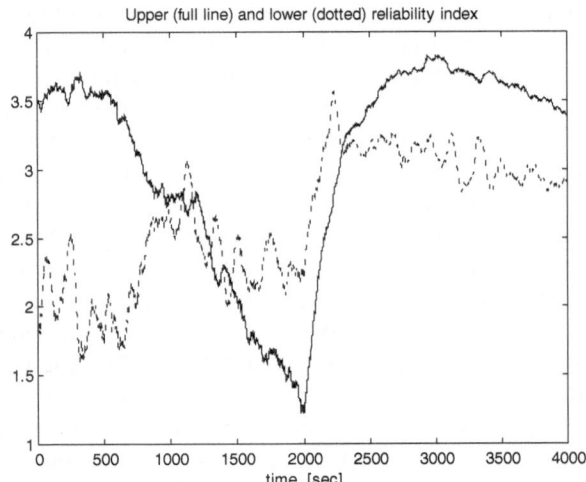

The vessel motion resulting from this control algorithm is shown in Fig. 5.18. The resulting set-point when activating the control action is located much closer to the well-head (at a position of −3.5 m initially and subsequently at a position of −7 m) than for the previous cases. The final position of the set-point even has a different sign than before (with the final position at +1.5 m).

Application of the symmetric objective function is next considered. The varying mean values of the angles are shown in Fig. 5.19. The shapes of these curves have some resemblance to those for the case with fixed weighting (where the weight for the bottom angle was 5 times that for the top angle). This is due to the fact that the ratio between the variances for the bottom versus the top angle for the present case is 1–4. This implies that the weight of the bottom angle becomes four times that

Fig. 5.18 Vessel position as function of time for positioning scheme based directly on simplified reliability indices (unsymmetric objective function). *Bottom angle* $\theta_{cr} = 2.5°$, $\beta_{cr} = 2$. *Top angle* $\theta_{cr} = 5°$, $\beta_{cr} = 1$ (From [9])

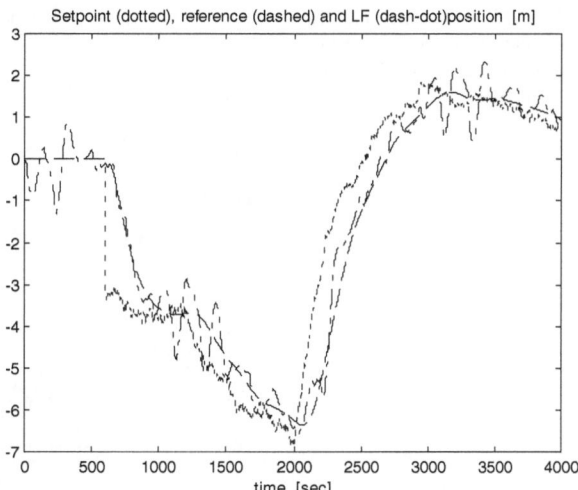

Setpoint (dotted), reference (dashed) and LF (dash-dot)position [m]

time [sec]

Fig. 5.19 Mean values of *top* and *bottom angles* for positioning scheme based directly on simplified reliability indices (unsymmetric objective function). *Bottom angle* $\theta_{cr} = 2.5°$, $\beta_{cr} = 2$. *Top angle* $\theta_{cr} = 5°$, $\beta_{cr} = 1.0$ (From [9])

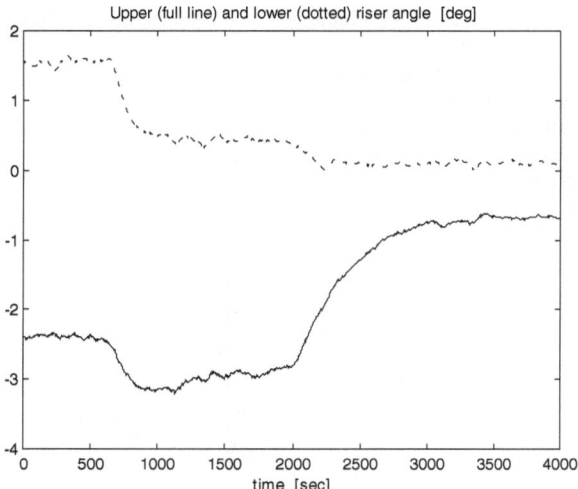

Upper (full line) and lower (dotted) riser angle [deg]

time [sec]

for the top angle since the weights are inversely proportional to the respective variances. Accordingly, the present results bear some similarity to the fixed-weighting case.

The reliability indices corresponding to this strategy are shown in Fig. 5.20. It is seen that the index for the top angle crosses just below 1.0 which is set as the critical value. The reliability index for the bottom angle stays above the critical value of 2.0. Again, these results are quite similar to the case with fixed weighting.

The vessel motion resulting from this control algorithm is shown in Fig. 5.21. The motion of the surface vessel is obviously also quite similar to the fixed weighting case. The initial optimal position is located quite close to that case, i.e. at about -16 m. The final position is at -4 m which is also very similar.

Fig. 5.20 Reliability indices as functions of time based on the mean values in Fig. 5.19 (From [9])

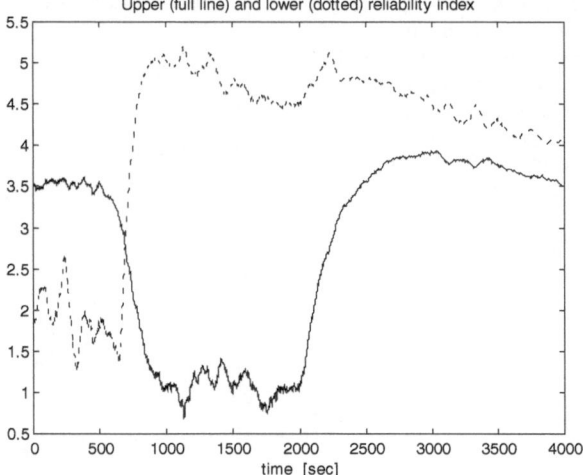

Fig. 5.21 Vessel position as function of time for positioning scheme based directly on simplified reliability indices (symmetric objective function). *Bottom angle* $\theta_{cr} = 2.5°$, $\beta_{cr} = 1$. *Top angle* $\theta_{cr} = 5°$, $\beta_{cr} = 2.0$ (From [9])

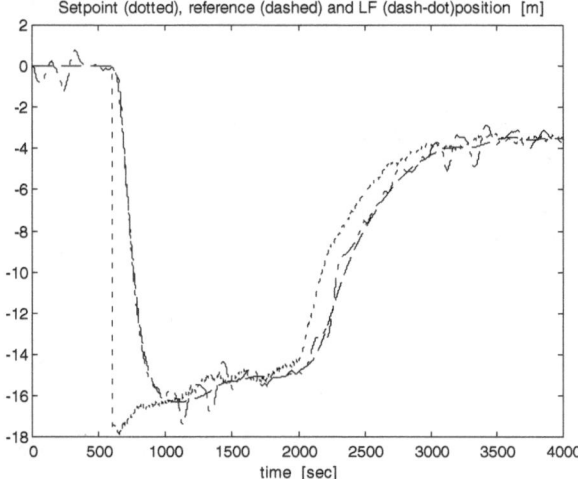

It is quite instructive to consider the shape of the objective function (or loss function) for the symmetric case, see Fig. 5.22. This function depends on several parameters. The values of these are given in the figure. The mean value of the bottom angle is 1° and the mean value of the top angle is 2°. The standard deviation of the bottom and the top angle are taken to be proportional to the same scaling factor (which is to be multiplied by 0.4° for the top angle and 0.1° for the bottom angle). The parabolic variation of the objective function for varying incremental position of the vessel is clearly observed. The minima of the collection of all such parabolas are located along the same straight line which intersects the "incremental position axis" at 90° at the optimum value which is equal to 15.3 m for the present case.

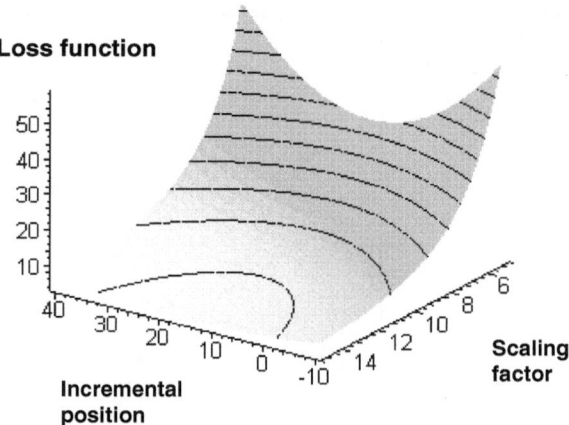

Fig. 5.22 Loss function for varying incremental vessel position and standard deviation scaling factor. *Top angle* coefficient $= -0.0420$, *Bottom angle* coefficient $= -0.0688$, *Top angle* mean value $= 2.0°$, *Bottom angle* mean value $= 1.0°$, *Top angle* standard deviation $= 0.4 *$ Scaling factor, *Bottom angle* standard deviation $= 0.1*$Scaling factor. (*Critical top angle* $= 5.0°$, *Critical bottom angle* $= 2.5°$, Critical reliability index *top angle* $= 1.0$, Critical reliability index *bottom angle* $= 2.0$) (From [9])

If only a single riser angle is to be minimized, the algorithm is much simpler as discussed above. The resulting mean angle time series for the unsymmetric objective function are given in Fig. 5.23. After activation of the control algorithm the bottom angle is first reduced to 1.3 and subsequently to 0.8°. The reason for the non-zero value of the angle is that the achieved mean values are sufficient to maintain the target reliability level.

Fig. 5.23 Mean values of top and bottom angles for positioning scheme based on control of bottom angle only (Specialized control algorithm based on unsymmetric loss function) (From [9])

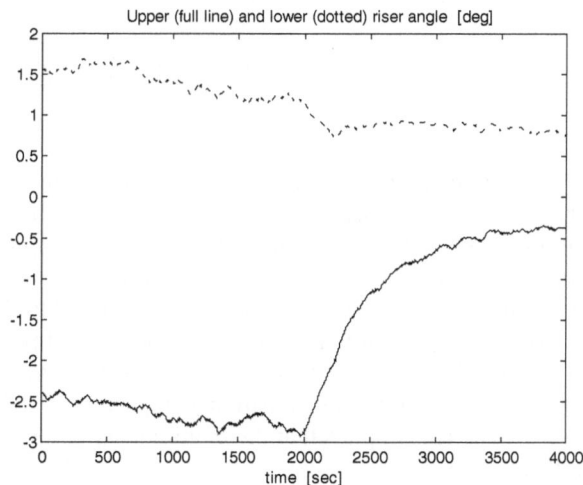

Fig. 5.24 Reliability indices
as function of time based on
the mean values in Fig. 5.23
(From [9])

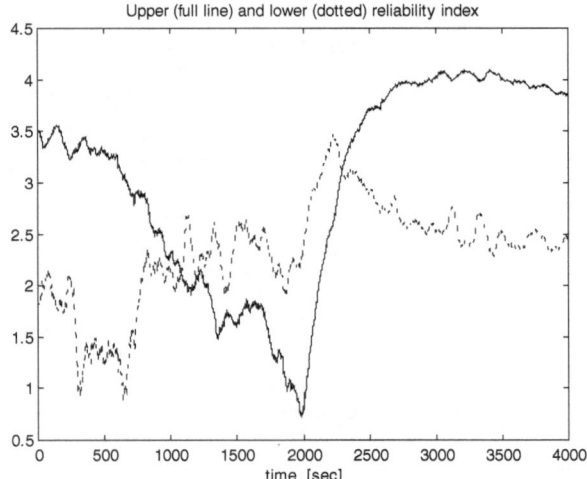

Fig. 5.25 Vessel position as
function of time for
positioning scheme based on
control of bottom angle
(Specialized control
algorithm based on
unsymmetric loss function)
(From [9])

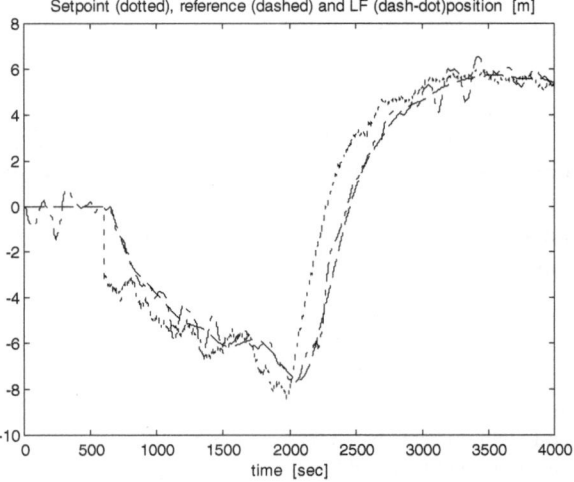

This is seen more clearly from the reliability indices in Fig. 5.24. The index for
the bottom angle stays above the critical value of 2.0 after the positioning algorithm has been activated. The index for the top angle is shown for curiosity based
on the critical value applied above. Not unexpectedly, this index is very close to
zero during part of the time since criteria related to the top angle are not included.

The corresponding time variation of the vessel position is depicted in Fig. 5.25.
It is seen that the vessel motion is much more symmetric than for the previous
case, with the minimum position being around -7 and the maximum position
around $+6$. This implies that the required vessel motion relative to the well-head is
rather limited.

As a main observation from the present example, it is found that on-line computation of the simplified structural reliability index is quite straightforward and provides response levels that correspond to the target values which are specified. Other studies related to control of riser angles based on different types of objective functions are found e.g. in [13, 14].

5.4.2 Example 2: Position Mooring of Floating Vessel Based on Reliability Index Criteria

The next example is based on [15] where the delta-index is implemented as an integral part of the control loop for the purpose of position mooring, which corresponds to dynamic positioning of a surface vessel moored to the seabed via a turret-based spread mooring system. While the mooring system keeps the surface vessel in place most of the time, thruster assistance is needed in severe weather conditions in order to reduce the probability of mooring line failure. Traditionally, this is done by keeping the vessel within a specified area which is predetermined, i.e. computed outside the control loop. The feasibility of the present controller is verified by laboratory experiments.

Implementation of structural reliability criteria as part of an extended LQG scheme was considered in Sect. 5.3. A completely interactive algorithm based on the instantaneous value of the delta index was not attempted in that case. Instead, specific vessel offset limits were set a priori for both initial and full thruster activation. These offset values were computed based on corresponding specified (target) values of the associated index.

In the following, a control scheme based completely on the instantaneous value of the so-called delta index is presented. The control scheme is described in the following. This controller design deviates from the PID algorithm as it is based on back-stepping techniques.

The delta-index is expressed in terms of the tension for mooring line number k as

$$\delta_k(t) = \frac{T_{b,k} - T_k(r_k(t)) - g\sigma_k}{\sigma_{b,k}} \text{ for k} = 1, \ldots, q \qquad (5.47)$$

where q is the number of mooring lines; $T_{b,k}$ is the mean breaking strength of mooring line k; $T_k(r_k(t))$ is the slowly-varying tension (i.e. static tension plus tension induced by wind and slow-drift forces); $r_k(t)$ is the time-varying position of the attachment point of the mooring line at the floater; σ_k is the standard deviation of the wave-induced dynamic tension, g is a "gust-effect" scaling factor, and $\sigma_{b,k}$ is the standard deviation of the mean breaking strength. The failure probability (corresponding to a given reference duration) for the critical mooring line is furthermore expressed in terms of the delta-index as $p_f = \Phi(-\delta)$.

The controller objectives are: (1) To ensure structural integrity of the mooring system by keeping the reliability index of all mooring cables above a preset safety limit; (2) To regulate the heading to a preset desired value, and; (3) To employ motion damping by limiting velocities in all three degrees of freedom.

We assume that both position and velocity is available for feedback, however we will need an estimate of the environmental force, $\mathbf{b}$, see Eq. (5.4) above. For this purpose, one of the designs in [16] is utilized:

$$\dot{\mathbf{v}} = -\mathbf{M}^{-1}\mathbf{D}\mathbf{v} - \mathbf{M}^{-1}\mathbf{g}(\boldsymbol{\eta}) + \mathbf{M}^{-1}\boldsymbol{\tau} + \mathbf{M}^{-1}\mathbf{J}^T(\psi)\hat{\mathbf{b}} - \mathbf{A}_m\tilde{\mathbf{v}}, \tag{5.48}$$

where

$$\tilde{\mathbf{v}} = \mathbf{v} - \hat{\mathbf{v}}, \tag{5.49}$$

and

$$\mathbf{A}_m = \mathbf{A}_0 - \lambda\mathbf{M}^{-T}\mathbf{M}^{-1}\mathbf{P}, \tag{5.50}$$

where $\lambda > 0$, $\mathbf{P}$ is the covariance matrix of the velocity vector, and $\mathbf{A}_0$ is a matrix satisfying

$$\mathbf{P}\mathbf{A}_0 + \mathbf{A}_0^T\mathbf{P} = -\mathbf{I}, \ \mathbf{P} = \mathbf{P}^T > \mathbf{0} \tag{5.51}$$

The update law for $\hat{\mathbf{b}}$ is given by

$$\dot{\hat{\mathbf{b}}} = \boldsymbol{\Gamma}\mathbf{M}^{-1}\mathbf{P}\tilde{\mathbf{v}}, \tag{5.52}$$

where $\boldsymbol{\Gamma} = \boldsymbol{\Gamma}^T > \mathbf{0}$. The error dynamics for the identifier becomes

$$\begin{aligned}\dot{\tilde{\mathbf{v}}} &= \left(\mathbf{A}_0 - \lambda\mathbf{M}^{-T}\mathbf{M}^{-1}\mathbf{P}\right)\tilde{\mathbf{v}} + \mathbf{M}^{-T}\mathbf{J}^T\tilde{\mathbf{b}}, \\ \dot{\tilde{\mathbf{b}}} &= -\boldsymbol{\Gamma}\mathbf{M}^{-1}\mathbf{P}\tilde{\mathbf{v}}\end{aligned} \tag{5.53}$$

The identifier error dynamics described by (5.53) are globally asymptotically stable (GAS). Thus, the following feedback controller can be designed via the back-stepping technique:

$$\boldsymbol{\tau} = \mathbf{M}\boldsymbol{\zeta} + \begin{bmatrix} \frac{T_j'}{\sigma_{bj}}\bar{\delta}_j\vartheta \\ -(\psi - \psi_s) \end{bmatrix} + \left(\mathbf{D} + \boldsymbol{\Lambda}\right)\begin{bmatrix} \bar{\delta}_j\gamma\vartheta \\ -\kappa(\psi - \psi_s) \end{bmatrix} - \boldsymbol{\Lambda} - \mathbf{J}^T(\psi)\hat{\mathbf{b}} + \mathbf{g}(\boldsymbol{\eta}) \tag{5.54}$$

where

$$\bar{\delta}_j = \min\{0, \delta_j - \delta_s\}, \tag{5.55}$$

with δ_j denoting the instantaneous value of the delta index while δ_s is the specified minimum permissible value. Furthermore

$$\boldsymbol{\zeta} = \begin{bmatrix} \gamma(\xi\mathbf{I} + \rho\bar{\delta}_j\mathbf{S}_2)\vartheta + \frac{\gamma\bar{\delta}_j}{r_j}(\mathbf{I} - \vartheta\vartheta^T)\mathbf{w} \\ -\kappa\rho \end{bmatrix}, \tag{5.56}$$

$$\vartheta = \mathbf{J}_2^T(\psi)\frac{(\mathbf{p} - \mathbf{p}_j)}{r_j}, \tag{5.57}$$

$$\xi = \left\{ array*20ll0, \delta_j > \delta_s - \frac{T_j'}{\sigma_{b,j}}\vartheta^T w, \delta_j \leq \delta_s, \right. \tag{5.58}$$

$$\mathbf{S}_2 = \begin{bmatrix} 0 & 1 \\ -1 & 0 \end{bmatrix}. \tag{5.59}$$

In the above, ρ is a constant with a value which is bounded by the maximum extension of mooring line number j, and ψ_s is the specified heading. The controller in Eqs. (5.54)–(5.59), in closed loop with the identifier in Eqs. (5.48)–(5.52) renders the set

$$\mathbf{M} = \left\{ (\boldsymbol{\eta}, \mathbf{v}, \tilde{\mathbf{v}}, \tilde{\mathbf{b}}) : \psi = \psi_s, \delta_j \geq \delta_s, \mathbf{v} = \tilde{\mathbf{v}} = \tilde{\mathbf{b}} = \mathbf{0} \right\}, \tag{5.60}$$

globally asymptotically stable (see [15] for a proof of this property).

Both numerical simulations and experimental tests were performed in order to verify the behavior of the present control scheme. The vessel used is called *CyberShip III* (CS3), see Fig. 5.26. CS3 is a scaled model (1:30) of an offshore supply vessel, and is equipped with 4 thrusters, three are fully rotatable and one is a fixed bow thruster. The main characteristics of the CS3 model, and the full scale version, are as shown in Table 5.1 (Note that the thrusters which are used on the model vessel are not correctly scaled relative to each other).

Fig. 5.26 Cybership III
(From [15])

Table 5.1 Main characteristics of CS3

	Model	Full Scale
Length over all	2.275 (m)	68.28 (m)
Length between perpendiculars	1.971 (m)	59.13 (m)
Breadth	0.437 (m)	13.11 (m)
Draught	0.153 (m)	4.59 (m)
Weight	74.2 (kg)	2.3×10^6 (kg)
Azimuth thrusters (3)	27 (W)	1,200 (kW)
Bow thruster	27 (W)	410 (kW)

The tests were performed in MCLab, which is a test basin specifically designed for control of marine vessels, with a moveable bridge, where the operator can supervise the experiments. In the experimental setup, a mooring cable was attached to the bow of CS3 and fastened to the basin. Only one mooring cable was used, which is sufficient since the controller only considers the most critical δ-index. The tension in the cable was measured with a force ring. The measurement of the tension was applied for calculation of the δ-index. To simulate the slowly varying forces such as current, wind and second order wave loads, a cord was tied to the aft of the vessel and attached weights were applied in order to drag it backwards. Changes of the weights were introduced instantly, representing a step in the environmental loads. Although this does not represent the actual transition between two weather conditions but rather an extreme case, it yielded important information about how the controller reacts to an abrupt increase in the environmental loading. The position measurement is based on application of four cameras detecting five light-balls on the vessel. These are seen in Fig. 5.26 and they are flashing at 50 Hz.

The results presented below are for incoming irregular waves with a significant wave height of 0.03 m, corresponding to a full-scale significant wave height of approximately 0.9 m. The δ -index is slightly affected as the controller is switched on at t = 0 min, and the heading is regulated to its desired value (see Fig. 5.27). The thruster force, shown in Fig. 5.28 has some variations. However, the variations are not very rapid and do not represent any high strain situation for the thrusters.

The environmental load is applied at t = 1.8 min and removed at t = 4 min. It is observed how the controller effectively acts to prevent the δ-index from going below the critical level of 4.0. At the time when the environmental load is removed the system is allowed to float freely again.

A comparison of optimal floater offset values which are obtained by application of different types of objective functions (loss functions) based on a simplified quasistatic response model is provided by Ref. [17]. Further comparisons between

Fig. 5.27 Time variation of δ and ψ (From [15])

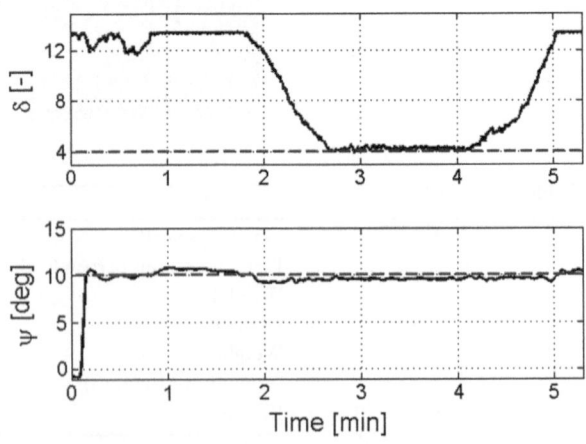

Fig. 5.28 Time variation of the applied thruster force for the x, y, and ψ components (From [15])

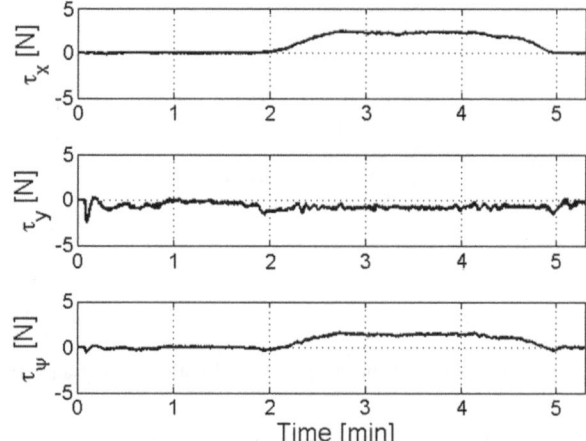

numerical simulations and experimental results for the implementation which was just present is found in [18]. A major benefit of the present control scheme is that global asymptotic stability (GAS) can be verified, which is a highly attractive property.

5.4.3 Further Examples of Application

In [19] an on-line reliability-based dynamic positioning scheme for a surface vessel moored to the seabed via a spread mooring system was considered. The objective function is of a similar quadratic type to the one which was applied for the riser angles presented in the example above. When environmental loads become high, the position mooring system applies thruster forces to protect the mooring lines. A new on-line position chasing algorithm is applied in order to protect all the mooring lines simultaneously. The tension levels in the mooring lines are included in the cost function where the degree of criticality for each line (expressed in terms of the δ-index) determines the individual priority weighting. With this strategy, external environmental effects are included directly without the need for predefined tabular settings of environmental conditions as in many earlier approaches. The delta-index is applied for derivation of weight factors to represent the dynamic influence of mooring line tension. Detailed simulations illustrate the features and advantages of the method and results are compared with those of an algorithm based on fixed weighting

A combination of reliability index criteria related to both mooring lines and riser system was investigated in [20]. Implementation of these criteria was studied in relation to a floating vessel at a water-depth of 1,000 m. A layout of the mooring system is shown in the left part of Fig. 5.29. The single riser is indicated by a vertical line in the right part of the figure.

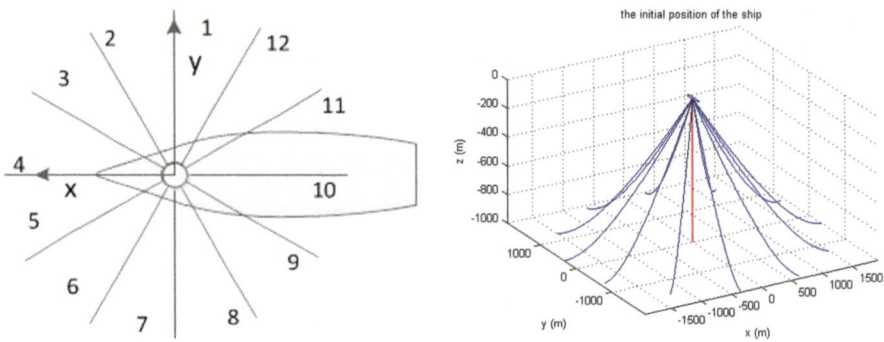

Fig. 5.29 Layout of the floating vessel with mooring and riser system (From [20])

The objective function for this case comprises δ-indices related both to the riser and the mooring lines:

$$L = w_t(\delta_{st} - \delta_t)^2 + w_b(\delta_{sb} - \delta_b)^2 + \sum_{i=1}^{p} w_i(\delta_{si} - \delta_i)^2 \qquad (5.61)$$

where subscript s designates the specified value, subscript t refers to riser top angle, subscript b refers to riser bottom angle, and subscript i refers to mooring line number i.

The optimal magnitude and direction of the position increment can then be obtained in a similar manner as for the riser example above by differentiation of the objective function. The resulting expressions for the magnitude and direction then become:

$$\Delta r_v = \frac{K_{11}^m \sin\theta + K_{12}^m \cos\theta - K_{11}^r \cos\theta}{K_{21}^m \sin^2\theta + 2K_{22}^m \sin\theta \cos\theta + K_{23}^m \cos^2\theta + K_{21}^r \cos^2\theta}$$

$$\theta_{op} = \mathrm{tg}^{-1}\frac{K_{11}^m K_{23}^m - K_{12}^m K_{22}^m + K_{22}^m K_{11}^r + K_{11}^m K_{21}^r}{K_{12}^m K_{21}^m - K_{11}^m K_{22}^m - K_{21}^m K_{11}^r} \qquad (5.62)$$

where K_{11}^m–K_{23}^m, K_{11}^r–K_{21}^r are constants that depend on the initial geometry of the mooring line configuration and the instantaneous values of the different indices. Superscript m here refers to the mooring line system, while superscript r refers to the riser.

The present reliability-based positioning scheme is applied in connection with fault tolerant control algorithms for mitigation of the consequences of an initial failure of a single mooring line. Figure 5.30 (upper left part) shows the x- and y-position (surge and sway directions) of the vessel. The upper right part of the figure shows the indices for the top and bottom riser angles. The lower part gives the corresponding indices for some of the mooring lines. In all cases the results *without* fault tolerant control (FTC) are shown by means of blue lines. The other branches give the results when FTC is applied in conjunction with the present repositioning scheme.

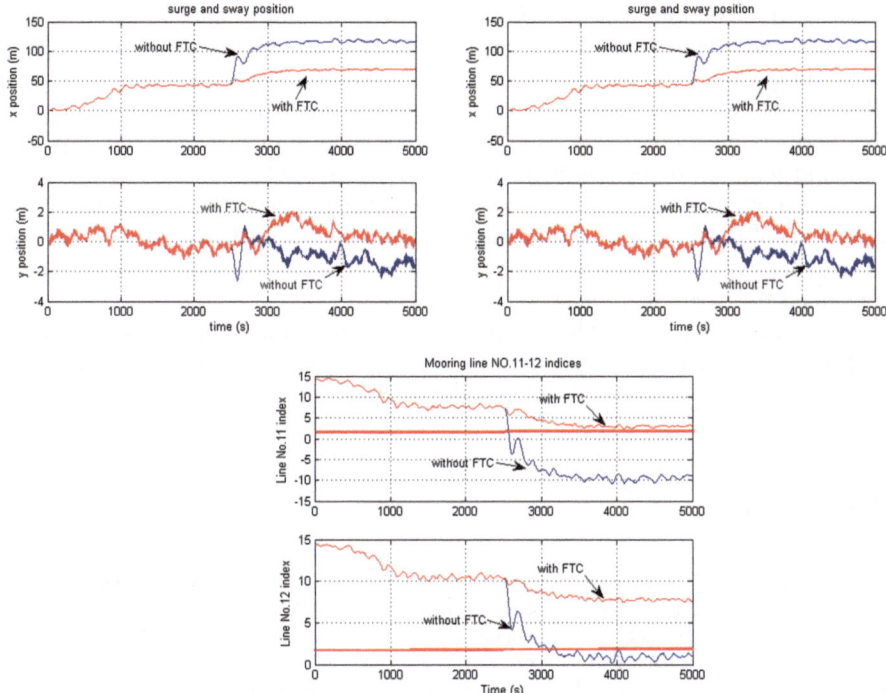

Fig. 5.30 Fault tolerant control. Floater position (*upper left*), riser angle indices (*upper right*) and mooring lines (*lower part*) (From [20])

From the results, it is clearly observed that the reliability indices are kept above the specified levels for all the relevant components when the FTC control action is introduced.

5.5 Concluding Remarks

Various approaches for incorporation of structural reliability measures were considered in relation to control procedures for floating vessels. Examples of three different categories of such procedures were discussed.

On-line monitoring in combination with an extended LQG algorithm was first considered. It was found that significant improvement of mooring line reliability was achieved by activation of the thrusters for only part of the time (i.e. around one-third).

As the next example, a PID control scheme based on continuous on-line evaluation of the structural reliability measure for reduction of riser angle magnitudes was investigated. Again, it was found that significant increase of the reliability level (i.e. reduction of the inherent failure probability) was achieved.

The third example was also concerned with on-line implementation of the associated structural reliability measure. The control scheme was of a slightly more complex type than for the previous examples, but this type of scheme allowed that Global Asymptotic Stability could be verified. Numerical simulations also verified by model tests showed that the algorithm provided the expected results in relation to position mooring of a floating vessel.

An example that included risers as well as mooring lines was also considered in combination with a fault tolerant control scheme. Again the behavior was found to be excellent as investigated by numerical simulations. Model have also been performed for verification of these results, [21].

As a main observation, there is clearly much room for further development of control schemes based on structural reliability criteria both within the areas that presently are investigated as well as similar applications within other fields.

References

1. Fossen, T. I. (2002). *Marine control systems*. Trondheim, Norway: Marine Cybernetics.
2. Sørensen, A. J., Lindegaard, K. P., & Hansen, E. D. D. (2002). Locally multiobjective H_2 and H_1 control of large-scale interconnected marine structures. *Proceedings of the 41st IEEE Conference on Decision and Control, Las Vegas, NV, US*.
3. Aamo, O. M., & Fossen, T. I. (2001). Finite element modelling of moored vessels. *Mathematical and Computer Modelling of Dynamical Systems, 7*(1), 47–75.
4. Berntsen, P. I. B., Leira, B. J., Aamo, O. M., & Sørensen, A. J. (2004, June 20–25). Structural reliability criteria for control of large-scale interconnected marine structures. *Proceedings of 23rd OMAE, Vancouver, Canada*.
5. Sørensen, A. J. (2002). *Marine cybernetics-modelling and control*. Trondheim: NTNU.
6. Sørensen, A. J., Sagatun, S. I., & Fossen,T. I. (1996). Design of a dynamic positioning system using model-based control. *Journal of Control Engineering Practice, 4*(3), 359–368.
7. Sørensen, A. J., & Strand, J. P. (2000). Positioning of small-waterplane-area marine constructions with roll and pitch damping. *Journal of Control Engineering Practice, 2*, 205–213.
8. Berntsen, P. I. B., Aamo, O. M., Leira, B. J., & Sørensen, A. J. (2004). Structural reliability-based control of moored interconnected structures. *Control Engineering Practice*, 2006, 10.1016/j.conengprac.2006.03.004.
9. Leira, B. J., Sørensen, A. J., & Larsen, C. M. (2004). A reliability- based control algorithm for dynamic positioning of floating vessels. *Structural Safety, Elsevier Ltd., 26*, 1–28.
10. Leira, B. J., Sørensen, A. J., Larsen, C. M. (2002). Reliability-based schemes for control of riser response and dynamic positioning of floating vessels. Paper 28436, *Proceeding of OMAE 2002, Oslo, Norway*.
11. Leira, B. J., & Sørensen, A. (2005). Structural reliability criteria for on-line control of marine risers. *Proceedings of Eurodyn 2005, Paris, France*.
12. Chen, Q. (2001). Analysis and control of riser angles. Master Thesis, Department of Marine Structures, Faculty of Marine Technology, NTNU, Trondheim, Norway.
13. Imakita, A., Tanaka, S., Amitani, Y., & Takagawa, S. (2000). Intelligent riser angle control DPS. *Proceedings of ETCE/OMAE2000 Joint Conference*, OMAE2000/OSU OFT-30001, New Orleans, US.
14. Nguyen, D. H., Nguyen, D. T., Quek, S. T., & Sørensen, A. J. (2010). Control of marine riser end angles by position mooring. *Control Engineering Practice, 18*, 1010–1020.

15. Berntsen, P. I. B., Aamo, O. M., & Leira, B. J. (2009). Ensuring mooring line integrity by dynamic positioning: Controller design and experimental tests. *Automatica, 45*, 1285–1290.
16. Krstic, M., Kanellakopoulos, I., & Kokotovic, P. (1995). *Nonlinear and adaptive control design*. Wiley, New York.
17. Leira, B. J., Berntsen, P. I. B., & Aamo, O. M. (2008). Station-keeping of moored vessels by reliability-based optimization. *Probabilistic Engineering Mechanics, 23*, 246–253.
18. Berntsen, P. I. B (2008). Structural reliability based position mooring. Phd Thesis, 2008:62, Department of Marine Technology, NTNU, Trondheim, Norway.
19. Fang, S., Blanke, M., & Leira, B. J. (2010). Optimal set-point chasing of position moored vessel. *Proceedings of OMAE 2010, Shanghai, China*.
20. Fang, S., Leira, B. J., & Blanke, M. (2011). Reliability-based dynamic positioning of floating vessels with riser and mooring system. *Proceedings of MARINE 2011, Lisbon, Portugal*.
21. Fang, S. (2012). Fault tolerant position mooring control based on structural reliability. Phd. Thesis, Department of Marine Technology, NTNU, 2012:156.

Chapter 6
Conclusions

Methods for incorporation of structural reliability measures as an intrinsic part of active control algorithms was elaborated in the present text. Three different categories of such procedures were presented and applied to a simplified example. Different and more realistic applications of the different categories of control schemes were next presented.

The first category of such procedures corresponds to pre-calibration of active control parameters which are associated with "classical algorithms". This was illustrated in connection with loss functions associated with Linear Quadratic Gaussian (LQG) control. An alternative loss function which incorporates the cost associated with structural failure due to overload was introduced for the purpose of pre-calibration. The optimal value of the "control factor" based on the cost of structural system failure was identified. Calibration of the coefficients for a pure LQG control scheme in order to comply with the associated optimal point was subsequently outlined.

The second category corresponds to activation of control energy based on monitoring the value of the structural reliability measure. This activation starts when the value of this reliability measure falls below a pre-defined threshold value. The scheme was first illustrated in connection with a simplified example and subsequently in connection with a more realistic application. For the latter example, continuous monitoring of the index was combined with an extended LQG algorithm for thruster-based position control of a floating system. It was found that significant improvement of mooring line reliability was achieved by activation of the thrusters for relevant time intervals of limited duration.

The third category of control algorithms is based on continuous on-line evaluation of the required control magnitude based on the structural reliability measure. After having re-visited the simplified example, two additional and more complex applications were addressed. First, a PID control scheme based on continuous on-line evaluation of the structural reliability measure for reduction of riser angle magnitudes was investigated. Again, it was found that significant increase of the reliability level (i.e. reduction of the inherent failure probability) was achieved.

B. J. Leira, *Optimal Stochastic Control Schemes Within a Structural Reliability Framework*, SpringerBriefs in Statistics, DOI: 10.1007/978-3-319-01405-0_6, © The Author(s) 2013

The second example from this "on-line category" was also concerned with dynamic positioning of a floating vessel. A reliability measures related to the mooring line tension was applied. The control scheme was of a slightly more complex type than for the previous examples, but this type of scheme allowed that Global Asymptotic Stability could be verified. Numerical simulations as well as model tests showed that the algorithm provided the expected results in relation to position mooring of a floating vessel. An extended example that included risers as well as mooring lines was also briefly outlined in combination with a fault tolerant control scheme.

As a main observation, there is clearly much room for further development of control schemes based on structural reliability criteria both within the present area (i.e. positioning schemes for floating vessels) as well as similar applications within related fields. It is intended that the present text may serve to point at some basic approaches to such developments.